Dilbarhon Fazilova

ANALYSIS OF TECTONICS OF TASHKENT REGION BY SATELLITE METHODS

Dilbarhon Fazilova

ANALYSIS OF TECTONICS OF TASHKENT REGION BY SATELLITE METHODS

ScienciaScripts

Imprint

Cover image: www.ingimage.com

This book is a translation from the original published under ISBN 978-620-7-45223-1.

Publisher:
Sciencia Scripts
is a trademark of
Dodo Books Indian Ocean Ltd. and OmniScriptum S.R.L publishing group

120 High Road, East Finchley, London, N2 9ED, United Kingdom
Str. Armeneasca 28/1, office 1, Chisinau MD-2012, Republic of Moldova, Europe
Printed at: see last page
ISBN: 978-620-8-29609-4

Contents

INTRODUCTION

At present, the study of the stress-strain state of the Earth's crust, leading to the destruction of the lithosphere and the formation of tectonic plates on various spatial and temporal scales, is one of the most urgent physical problems solved by space geodesy methods. Monitoring of geodynamic processes on various temporal and spatial scales using satellite methods is due to high operational efficiency, economic efficiency, objectivity and reliability of results, as well as the possibility of detecting phenomena and processes inaccessible by other methods. From this point of view, the task of determining the parameters of geodynamic processes on the basis of satellite data becomes relevant.

At present, the world is improving the technology of satellite data processing to provide an adequate assessment of the level of modern geodynamic activity of the environment and reliable identification of the results of observations of deformation process parameters. Targeted scientific research is being conducted in this area, including: monitoring of the movement of tectonic plates, analysis of their velocities and directions using the global navigation satellite system (GNSS), study of changes in the height of the earth's surface, analysis and forecasting of various geological phenomena on the basis of automated processing of multispectral and radar satellite images, development of methods for assessing the stability of zones with increased seismic activity, analysis of the impact on tectonic activity, and analysis of the influence of satellite data on the earth's surface.

In the Republic of Uzbekistan much attention is paid to fundamental studies of deformation processes and their changes in time, caused by both tectonic and anthropogenic loads, on the basis of satellite measurements.

Significant results have also been achieved in the applied area of creating an instrumental base for reliable identification of the nature of the stress-strain state of the Earth's crust through joint processing of satellite images and new measurement technologies such as GNSS. However, on the territory of Uzbekistan, where active tectonic processes are observed, geophysical methods of research are currently predominantly used. It is important to note that there is no comprehensive approach combining geophysical methods with spatial satellite methods, which limits the possibilities of comprehensive research in this territory. In order to implement the tasks outlined in the "Strategy for the Development of New Uzbekistan for 2022-2026",[1] it is necessary to increase the efficiency of geoinformation support through the development of theoretical foundations for further introduction of modern innovative satellite technologies and transition to satellite-based automatic methods for analysing deformation processes.

This research work to a certain extent corresponds to the tasks outlined in the Decree of the President of the Republic of Uzbekistan No. P11-429 "On additional measures for further development of the space industry" dated 23.11.2022, in the Decree of the President of the Republic of Uzbekistan No. UP-144 "On measures for further improvement of the seismic safety system of the Republic of Uzbekistan" dated 30 May 2022, as well as in other regulatory and legal documents related to this area of activity.

This study was carried out in accordance with the priority directions of science and technology development of the Republic of Uzbekistan: IV. "Development of informatisation and information and communication technologies" and VIII. "Earth Sciences (geology, geophysics, seismology and processing of mineral raw materials)".

The study was carried out in accordance with the plans of research works of the Astronomical Institute, within the framework of fundamental research of the Space Research Laboratory of the Astronomical Institute of the Academy of Sciences of the Republic of Uzbekistan on the theme "Study of the regional stress state of the Earth's crust of the Republic of Uzbekistan by modern satellite methods" (2020-2023).

The aim of the study is to analyse tectonic processes in the Tashkent region using satellite methods for a deeper understanding of the crustal dynamics in this region.

The objectives of the study are:

improvement of the method of automatic joint processing of multispectral satellite data and digital elevation models for reliable detection of tectonic structures;

estimation of accuracy of digital models of normal heights from radar satellite measurements;

---

[1] Presidential Decree No. UP-60 "On the Strategy for the Development of New Uzbekistan for 2022-2026" dated 28 January 2022.

determination of the current spatial velocities of the points of the study area using the data of the global navigation satellite system;
interpretation of the dynamics and determination of the physical nature of lineament systems in the study area.
The object of the study are lineament structures and maps obtained from optical data (Landsat 8), digital elevation models (SRTM, ASTER, ALOS) and geological and geophysical data on the territory of the Tashkent region.
The subject of the study is satellite analysis methods (GNSS, multilevel automated lineament analysis) aimed at identifying and studying changes in geodynamics and crustal structure within the Tashkent region.
Research methods include the use of accurate geodetic techniques to measure displacements and deformations of the earth's surface, multi-level automated lineament analysis.
The scientific novelty of the study is as follows:
geoid model of the territory of Tashkent geodynamic polygon was built;
the catalogue of velocities of points in the Tashkent region was calculated;
it was revealed that the dynamics of changes in lineament structures and fracturing of the earth's surface is related to hydrological and seismic processes, as well as to the anthropogenic load throughout the study area.
The practical results of the study are as follows:
local geoid of the Tashkent geodynamic polygon was created;
local digital model of normal heights of Tashkent geodynamic polygon was created;
An improved algorithm for automated analysis of lineament structures to identify earthquake precursors has been developed;
values of horizontal and vertical velocities of reference points were obtained from the global navigation satellite system data.
Reliability of the study results is justified by long-term monitoring, a variety of analysis methods (including GNSS-geodesy, automated interpretation of satellite images and modelling) and comparison with independent data, providing reliability and accuracy of conclusions about tectonic processes in the Tashkent region.
The scientific significance of the research results lies in the possibility of using the developed methods and recommendations to describe modern deformations, including those of anthropogenic nature, of the Earth's crust

of the Tashkent region.
Practical significance of the results of the study is that they can be applied in the creation of an automated system for monitoring tectonic processes, including earthquakes and in the refinement of the system of normal heights of the Republic of Uzbekistan.
The scope and structure of the monograph. The work consists of an introduction, four chapters, a conclusion, and a list of references. The volume of the monograph is 109 pages.

CHAPTER I

GLOBAL NAVIGATION SATELLITE SYSTEMS FOR OBSERVATION OF GEODYNAMIC PROCESSES

1.1 Regional geodynamic studies by satellite methods

Uzbekistan, covering latitudes from 37° 11' to 45° 36' N and longitudes from 56° to 73° 10' E, lies in the transition zone from the Tien Shan mountain ranges to the Turan Platform. This area is a point of interaction between several major lithospheric plates, including the European, Asian, Iranian, Indian and Chinese plates. Studies conducted by specialists of the Vernadsky State Geological Museum of the Russian Academy of Sciences, such as Y.G. Gatinsky and D.V. Rundquist, using GNSS measurements, show that the Euro-Asian continent is not a homogeneous tectonic block. Rather, it consists of a solid North Eurasian plate, surrounded along its southeastern and eastern boundaries by a zone including several dozens of microplates [1; p.3-20]. Studies of structural and tectonic features of the Turan plate note the reactivation of faults about 200 km west of the Tien Shan. Late Palaeozoic faults show reactivation, indicating ongoing deformation of the Turan plate. These results are obtained from the works of Ulomov V.I. [2; pp. 14-30, 3; pp. 18-28]. In addition, active involvement of the Turan plate and the Kazakh shield in the processes of tectonic activation of the region was found [4; p.91-98]. The Central Tien Shan, especially the area west of the Talas-Fergana fault, shows pronounced counterclockwise rotational movements. The rates of horizontal movements vary from 4-5 to 0.5 mm/year and rapidly decrease in the north-west and west directions. The high rate of crustal displacement between the Talaso-Fergana and East Fergana faults is noticeable [2; p.14-30].

At the end of the last century, scientists of the Institute of Seismology of the Academy of Sciences of Uzbekistan, in particular, A.R. Yarmukhamedov and D.H. Yakubov, created a map of young and modern movements of Uzbekistan at a scale of 1:1000000. This map was developed on the basis of compilation and own geological and geomorphological data, including information from geodetic measurements made at geodynamic polygons of Uzbekistan. The map of

Eastern Uzbekistan on a scale of 1:500000 was also compiled by L.A. Khamidov [5; p.88-91].

Maps of modern horizontal and vertical motions based on satellite measurements have been compiled for the territory of Uzbekistan [6; pp. 1-130, 7; pp. 67-70, 8; pp. 168-171, 9; pp. 228-233]. Studies aimed at real-time estimation of displacements are being carried out, and a methodology for predicting earthquakes and their relationship with groundwater is being developed [10; p. 29-35]. Studies of the relationship between spatial and temporal trends of modern earth surface movements and seismicity have shown that the Western Tien Shan is a complex geodynamic system [6; p. 1-130]. Using the catalogue of earthquakes the reconstruction of the fields of the modern stress state of the Earth's crust in Uzbekistan was carried out. The method of cataclastic analysis of rupture displacements was applied (Fig.1.2). This approach provides a more accurate representation of deformations in the crust and can be used for more effective forecasting and management of geological risks in the region (Fig. 1.1) [11; p. 28-52]. As a result of the stress reconstruction it was obtained that for the Pritashkent region the near-latitudinal direction of the main axis of the maximum compression stress o3 was revealed. [11; c. 28-52]. Such works are carried out only on the basis of ground measurements obtained from measurements at seismic stations and geological methods. This, in turn, imposes a limitation in spatial coverage and leads to the relevance of using satellite data.

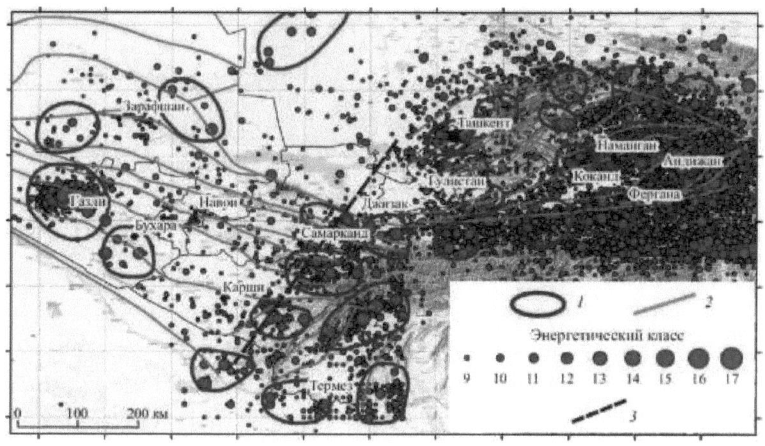

1 - focal zones; 2 - active faults; 3 - West Tien Shan lineament. Circles -

epicentres of earthquakes

Fig.1.1 Map of epicentres of strong earthquakes (M = 2.5-7.5) and faults [11;

c. 31]

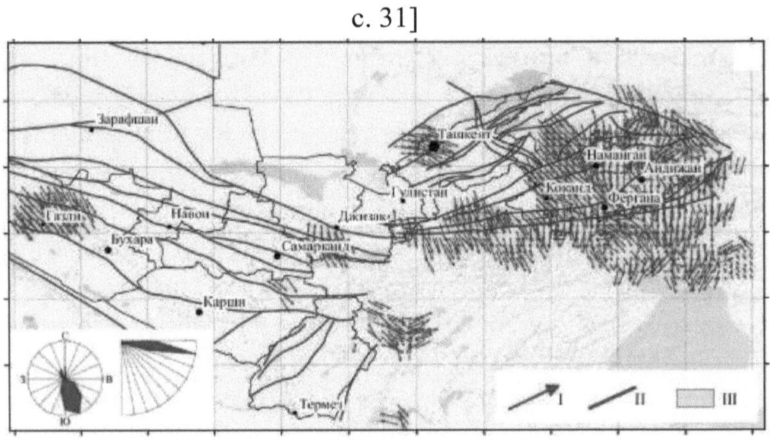

I - axes of minimum principal stress (maximum compression); II - active faults; III - seismogenerating zones: *1* - Talaso-Fergana; *2* - Chatkal-Atoynak; *3* - North Fergana; *4* - Namangan; *5* - Andijan; *6* - South Fergana; *7* - Kurshab; *8* - Taldysu; *9* - Chatkal

; *10* - Sandalash; *11* - Angren; *12* - Pskemsko-Tashkent; *13* - Nurekata; *14* - Lyangar; *15* - Ugam-Karjantau; *16* - Mogoltau-Pistalitau; 17 - Besapano-Severo-Nurata; *18* - Bukuntau; *19* - Severo-Tamdy; *20* - North-Kuljuktau-Turkestan; *21* - South-Auminzatau-Aktau; *22* - Zarafshan; *23* - Predkizilkum; *24* - South-Tian-Shan; *25* - Bukhara; *26* - Sultanuizdag; *27* - Gissar-Kokshaal; *28* - Kyzyl Darya-Liangar; *29* - Baysun-Kugitang; *30* - Surkhantau-Sherabad-Kelif; *31* - Babatag-Keykitau

; *31* - Babatag-Keykitau.

Fig.1.2. Projection of the axis of maximum compression (o3) in earthquakes (M <

4.5), sample N > 6, radius R < 30 km [11; p. 40]

The study of tectonic faults using geological methods has a long history and has yielded good results. At the same time, the periods between seismic events, which are the time of stress accumulation, are not sufficiently studied. Geological methods cannot do this due to their precise limitations, and geodetic survey by classical methods due to low velocities on most of the Tien Shan faults and the presence of distributed deformation between them gives large errors and requires considerable observation time.

Tectonic constructions are being improved using space geodesy techniques. Remote sensing, with its ability to provide synoptic and recurrent coverage, has in recent years become a key powerful tool for assessing and monitoring the dynamic activity of the Earth, including active tectonics and changes in surface topography in space and time. Studies by modern space geodesy methods, in particular GNSS, of the kinematics of this region and deformation distribution are devoted to the works of Abdurakhmanov K.E., Abdullabekov K.N., Khamidov L.A., Fazilova D.Sh., Ergeshov I.M. and others. [12-16].

The first studies using space geodesy methods in the region are related to quantitative assessments of modern earth surface movements on short time intervals. Since 1992, measurements have been carried out in the northern part of Central Asia on the international (Germany, Kyrgyzstan, Uzbekistan, Tajikistan, USA) network of GNSS observations CATS (Central Asia Tectonics Sciences) [12; 16; p.75-84]. The regional network covered the area from the northern to the eastern border of the Tajik depression, the Northern Pamir and Tarim basin over the Fergana basin, Chatkal and Tien Shan lines and the Kazakh platform. According to the results of network measurements, the rate of Tien Shan crust contraction in the north is ~15 mm/year [12; p. 450-453], in the south over 20 mm/year [13; p.42-45]. Modern approaches to emergency preparedness and disaster mitigation require obtaining information on geodynamic destructive phenomena as soon as possible and, if possible, in real time [17; p.10-15]. The Institute of Seismology of the Academy of Sciences of the Republic of Uzbekistan conducts research to identify morphokinematic indicators of modern geodynamics of the Western Tien-Shan based on the study of modern movements of the Earth's surface using GNSS measurements [18; p.49-54].

To solve the problem of predicting modern crustal movements and developing their physical foundations, it is necessary to study the changes in various rock properties under the action of loads or pressures in natural conditions. For this purpose, geodynamic polygons (GDPs) are created, where complex studies are carried out to investigate the peculiarities of spatial and temporal manifestation of earthquake precursors [19]. At present, there are also several dozens of geodynamic polygons of various purposes operating in the territory of Uzbekistan, most of them are designed to study modern crustal movements (MCM) in various structures and seismic zones to search for earthquake precursors. After the Tashkent 1966 and Tavaksay 1977 earthquakes, in order to search for precursors of strong earthquakes and to develop some issues of modern geodynamic activity of the Earth's crust, the Tashkent geodynamic polygon was organised, where, since 1978 up to the present time, fundamental research on the search for earthquake precursors has been carried out (Fig.1.6) [6; p.1-130, 20; p.5-8, 21; p.367-371].

1.2 Object of study

The Tashkent region in Uzbekistan is widely recognised for its growing industrial sector, comprising various industries such as coal mining, construction, power generation, mining and irrigation systems. The region is home to key industries including the Angren coal mine, Jigiristan open pit, Naugarzan and Apartak coal pits, and the Almalyk Mining and Metallurgical Combine, which plays an important role as the country's main producer of non-ferrous metals. The industrial centre of Almalyk, strategically located 50 km south-east of the capital Tashkent, serves as an important industrial hub. Located in a zone of significant tectonic activity, the region faces challenges exacerbated by the high level of water supply in the territory. It is worth noting that the Tashkent region occupies one of the two first places in the country in terms of seismic activity, as indicated by the ground acceleration map with maximum values up to 4.8 m/s^2 [22]. Moreover, a significant part of the Tashkent region is subjected to high compression stress, indicating negative dilatation [23]. Detailed studies focusing on displacements along fault systems have confirmed the predominance of the axis of maximum compression stress in the region close to the latitudinal axis [11]. These results emphasise the critical importance of careful monitoring and effective management of deformation processes in the region to reduce the risks associated with seismic events and potential hazards.

The Tashkent region is located in the Central (Middle) Tien Shan and is characterised by a complex and diverse tectonic structure. The study area is bounded to the south and south-east by the Chatkal and Kuramin Ranges, to the north and north-west by the Ugam Range and its branch - the Karjantau Range, and to the south by the Fergana Valley. In the east and north-east, the Talas Alatau Range is the border of the region. This region is characterised by a complex tectonic structure caused by intensive mountain formation and platform movements over many millions of years. The tectonic movements prevailing in the Middle Tien Shan include fold-fold-fold-elevated and shear-fold-fold-elevated mountain formation (Figure 1.3). The mountains here are formed by the collision of lithospheric plates, resulting in strong folding and crustal uplift [24]. The study area considered in this paper includes two of the nine major seismic active zones located in Uzbekistan and identified on the basis of seismological and seismotectonic data: Tashkent and Nurekatino-Angren. The region includes two large seismic active zones identified on the basis of seismological and seismotectonic data and capable of generating strong earthquakes: Tashkent and Nurekatino-Angren. The Tashkent seismic active zone has a length of about 240 km and a width of 30-40 km and stretches from NE to SW. Seismic manifestations within its boundaries are determined by the modern geodynamic activity of the Karzhantau fault in the northeastern part and the Tashkent flexural-fracture zone in the southwestern part. A number of strong earthquakes were also recorded within this zone during the instrumental period of observations: Burchmulin 1959 M = 5.9, Tashkent 1966 M = 5.3, Tavaksay 1977 M = 5.3, Nazarbek 1980 M = 5.1, Altyntyubinsk 1987 M = 5.1. The last activation of seismic active zone was manifested by earthquakes with magnitudes M = 4.8 and M = 4.6 in 2008 and 2010. The average period of recurrence of strong earthquakes in the Tashkent seismic active zone is 12-15 years. The second, Nurekati-Angren seismic active zone, has a length of about 300 km and a width of 20-30 km, extends from NE to SW. Seismic manifestations of this zone are caused by the dynamic influence of the Nurekata fault in the western part and the North and South Angren faults in the southern part. Koshtepinsk earthquake 1965 with M = 5.5; Bukin earthquake 1967 and Pskent earthquake 1970 with M = 5.0 are known within the zone. The last activation in this zone was manifested by the Tuyabuguz earthquake with M = 5.6 in 2013. The recurrence period of strong earthquakes in this zone is 12 years [25].

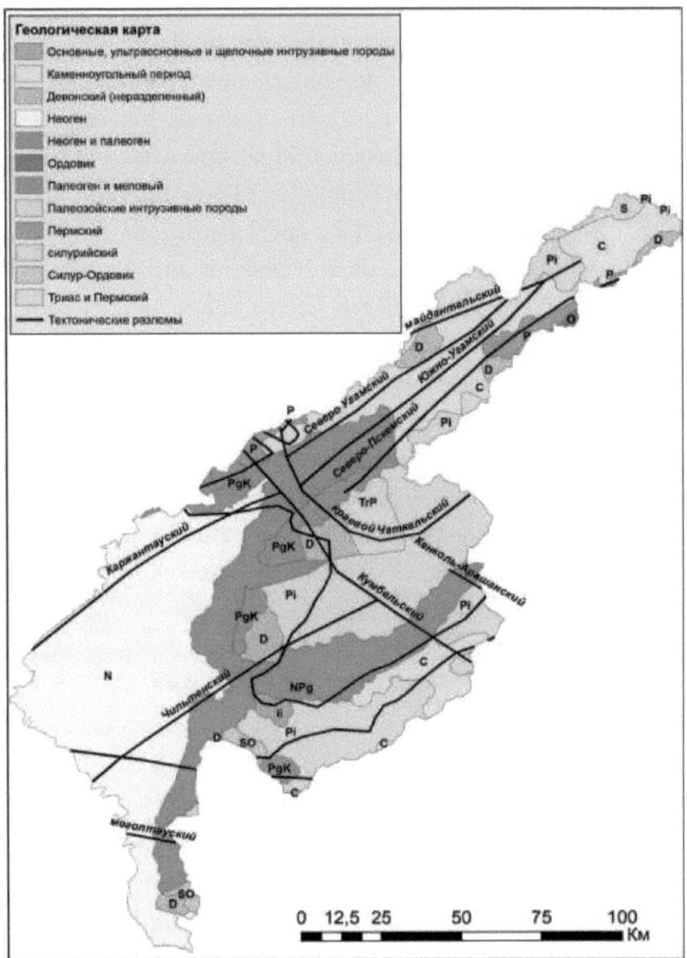

Fig.1.3. Geological map of the Tashkent region

Sh.H.Abdullaev et al. compiled the "Map of active rupture dislocations of the Republic of Uzbekistan" on a scale of 1:1000000. On the basis of available geological and geophysical facts on separate sites, the faults of Central Asia, including Uzbekistan, are genetically classified into deep (transcrustal), marginal (crustal) and newly formed (faults of the cover). Faults by structural-hierarchical subordination and geological-structural location

are subdivided into three ranks. The deep discontinuities of the 1st rank include the discontinuities separating geostructural areas: West-Tianshan,

South-Tianshan. The marginal discontinuities of the 2nd rank include the discontinuities separating the main structural elements within the geostructural areas (longitudinal) and secanting them (transverse). Longitudinal: North-Fergana, South-Fergana, Nurata, Turkestan , Amudarya, North-Bukantau, South-Kuljuktau. Transverse (secant): Ural-Oman, Persian-Balkhash, Tashkent-Dushanbe. All other faults (Fig. 1.4) are classified as marginal and newly formed faults of the 3rd rank [26].

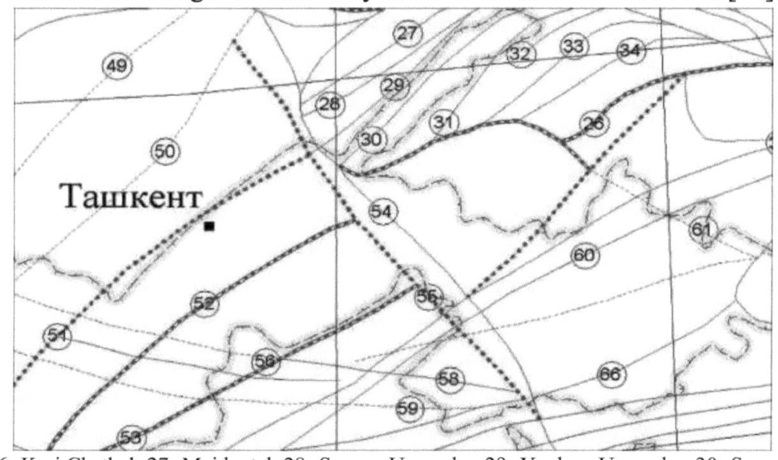

26- Krai Chatkal, 27- Majdantal, 28- Severo-Ugamsky, 29- Yuzhno-Ugamsky, 30- Severo-Pskemsky, 31- Yuzhno-Pskemsky, 32- Severo-Sandalashsky, 33- Yuzhno-Sandalashsky, 34- Chatkalsky, 49- Dzhausumsky, 50- Pritashkent, 51- Karzhantau, 52- Chiltensky, 53- Zhelezniy, 54- Kenkolsky, 55- Kumbel-Kokandsky, 56- Bashtavaksky, 58- Kurusay-Okurtau, 59- North-Mogoltau, 60- Prikuraminsky, 61- East-Fergana, 66- Tamdy-Karachatyrsky.

Fig.1.4. Schematic map of regional deep faults of the TGP and adjacent territory [26].

Charvak reservoir with an area of about 40 km^2 and a volume of 2 km^3 , designed mainly for hydropower purposes, is located in the TGP area. The height of the Charvak dam is 168m. During the annual cycle of the reservoir operation the volume of water in the reservoir changes from 100 to 2000 million m^3 , creating variable additional concentrated loads on the underlying rocks [27; p. 985-998]. This makes it possible to study local changes in the geomagnetic field caused by the dynamics of elastic stresses created Geomagnetic studies on modelling the processes of earthquake preparation in the Charvak reservoir area, conducted by the Institute of Seismology of the Academy of Sciences of the Republic of Uzbekistan, revealed that local anomalies of the geomagnetic field associated with the process of water volume change in the reservoir and local seismicity occur simultaneously by gravitational loading due to the mass of water [20-21].

1.3 Study of GNSS point offsets in the region

Space geodesy methods make it possible to determine horizontal movements of the Earth's crust with high accuracy. The range of values of modern movements varies from ~ 1 cm/year to 15 cm/year and is mainly the result of large deformations of tectonic plates at boundaries. The values of vertical crustal movements, on the other hand, are usually in the range of 1-10 mm / year, i.e. the magnitude is an order of magnitude smaller than horizontal movements. Therefore, they are more difficult to measure. Vertical movements are the result of various factors: deep tectonic processes responsible for the so-called slow modern movements of the Earth's crust and causing regular secular fluctuations of tectonic blocks or tectonic structures, seasonal ground fluctuations, anthropogenic influences, as well as some short-period movements, possibly also of tectonic origin, but not subject to regularities and rather random in nature. Non-periodic vertical movements caused by the above phenomena can be both local, regional and global. To date, they are poorly understood for high-precision modelling [34]. Separate studies were previously performed to determine the velocity field in the Tashkent region [2; 18; 12; 38].

Observations using a network of Global Navigation Satellite Systems (GNSS) points, the high accuracy and resolution of which allow to determine small changes in geocentric coordinates of ground points at short time intervals, are the basis for the development of national infrastructure of spatial data of the Republic of Uzbekistan. A GNSS geodetic network has been established in the country, the measurements of which are used to solve such problems as the construction of a national coordinate system, research of crustal deformations, and geoid determination [2832]. Measurements at more than 50 permanent points in the country are also becoming the basis for observations in the field of geosciences, in particular, the construction of a spatial model of Eurasian plate velocities in the region and obtaining information on geodynamic destructive phenomena in the shortest possible time in real time [33].

The region currently has a GNSS network of 12 stations including ALMA, ANGR, BAUD, BUKD, CHIR, GAZA, GULD, KELE, MTAL, SAYX, TASH and TOYT (Figure 1.5). For our study, we selected 6 measurement campaigns to analyse based on data availability. The duration of these measurement cycles, conducted from October 2018 to October 2020, was approximately 30 days each (Table 1.1.1). The stations are equipped with arc-shielded geodetic antennas and Leica/Trimble receivers, with the antennas' position angle maintained below 10°. Data are collected at approximately 30 second intervals.

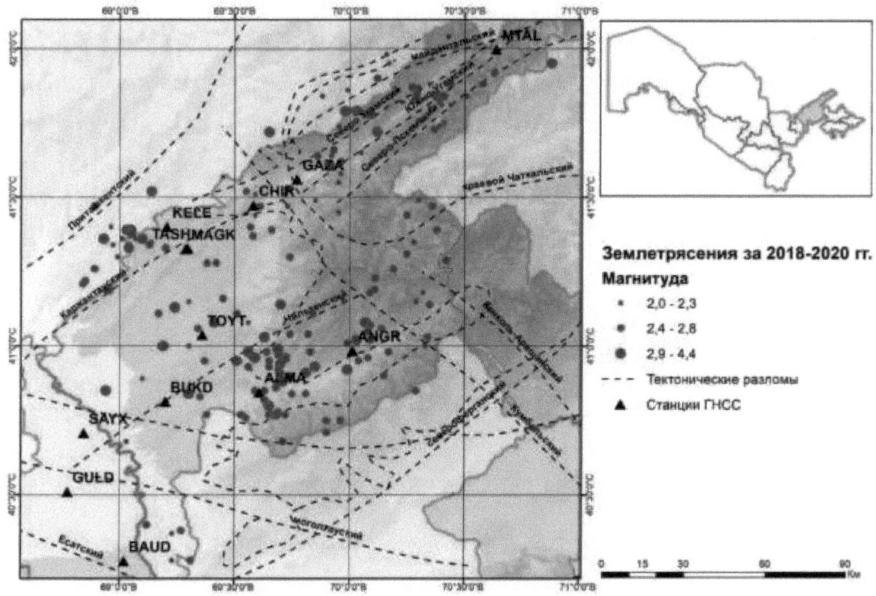

Fig.1.5. GNSS network of Tashkent region

Table 1.1.

Periods (days of months) of observations used for analysis

	January	February	March	April	July	August	September	October
2018	-	-		-	-	-	-	1-26
2019	10-31	1-28	1-31	1-30	-	-	29-30	1-26
2020	1-31	-	-	1-30	1-31	1-31	1-30	1-31

The measurements were processed in the GAMIT/GLOBK software of the Massachusetts Institute of Technology on the basis of the least squares method according to the recommendations and standards of the International Earth Rotation Service [35-36]. In order to obtain a stable solution of coordinates and velocities and to accurately link regional measurements to the international reference coordinate system ITRF2014 [37], in addition to international points (KITG, KIT3, TASH), 13 stations of the international GNSS service for geodynamics IGS (BJFS, BADG, LHAZ, URUM, PODG, KRTV, HYDE, IISC, CHUM, POL2, KAZA, KURG, TALA) were included in the processing. At the first stage of RINEX files processing in the GAMIT module, the quality of measurements was assessed using standard deviation values (RMS) for satellites and stations. The error range for the best network points is 4 mm to 5.3 mm and for the worst points is 8 mm to 12.4 mm. The ambiguities were resolved with 97 % for the wide band and 91 % for the narrow band, which confirms the noiseless pseudo-distance and indicate that the accuracy of the raw data is sufficient to obtain a reliable solution [35]. The normalised RMS is around 0.19 mm for all sessions. Figure 1.6 shows the velocity distribution of the region both relative to the Eurasian plate and the local displacements obtained by "plate stabilisation".

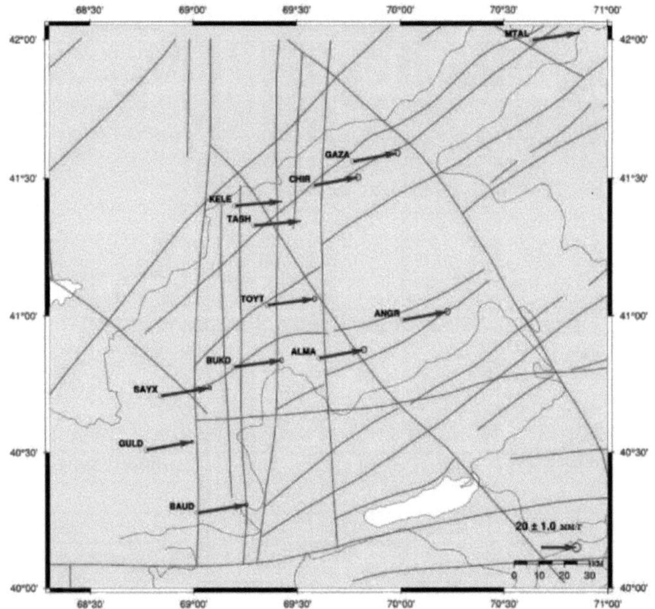

Fig. 1.6. Horizontal velocity field of the region relative to the Eurasian plate

Within the study area, all points indicate significant horizontal displacements with a 95% confidence level. The northeastern direction, corresponding to the Eurasia movement, of tectonic displacements ranges from 21 mm/y to 33 mm/y. Local motions, obviously having zonal characteristics due to tectonic faults of the region, reach values from 1.82

mm/y to 6.04 mm. The average velocity of the stations of the region is about 4 mm/y for the considered in the
work of the period from 2018 to 2020 (Figure 1.7).

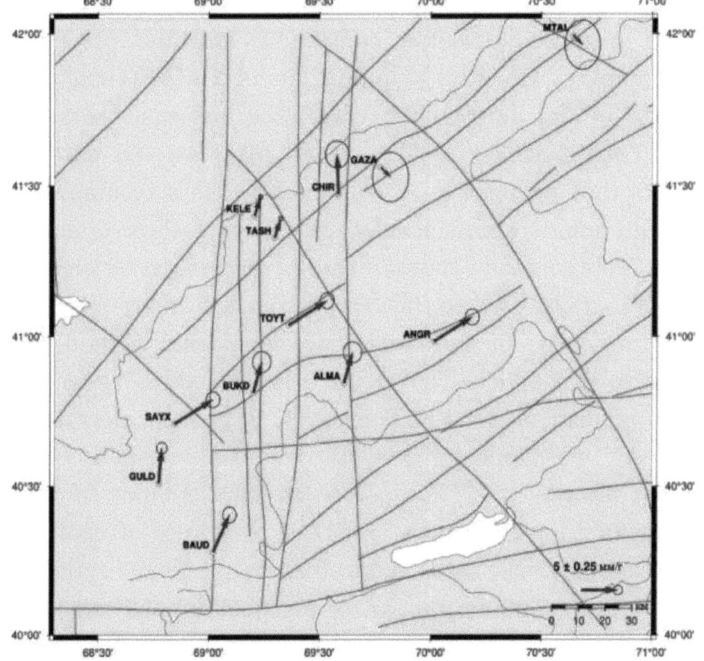

Fig. 1.7. Horizontal field of local displacements at "fixed Eurasia"

1.4 Construction of regular velocity field of Tashkent region based on interpolation of GNSS points data

The point velocities or the so-called velocity model obtained in the GAMIT/GLOBK programme allows describing the motion of the Eurasian plate in the study area. By "fixing the Eurasian plate" in the programme, it is possible to calculate displacements that determine the motion characteristics of local faults in the region or the effect of anthropogenic loads. But, to obtain a more informative model of the kinematics of the region, it is necessary to build a continuous velocity model by interpolating the obtained discrete values over the rest of the region. To refine "block" or "continuous" tectonic models, it is necessary to have a denser network of measurements and, in this case, interpolation of these discrete points to the whole area not covered by measurements to obtain a continuous velocity model is an important task. Various authors have used interpolation methods such as least squares method using the covariance function [39], nearest neighbour method, weighted inverse distance method [40], minimum curvature method [41], kriging [42], etc. However, these interpolation methods are time consuming and the process itself is relatively complex. Generic Mapping Tools GMT (Generic Mapping Tools) is one of the most widely used free and open source mapping software in Earth sciences with powerful mapping and data processing functions. For data processing, GMT features data screening, resampling, time series filtering, two-dimensional grid filtering, three-dimensional grid interpolation, polynomial approximation, regression analysis, etc. In addition, GMT can be used for fast and convenient interpolation and strain rate calculation [43]. In this paper, we used the method of coupled interpolation of two-dimensional velocity field vectors using the Green's function and implemented in GMT using the gpsgridder algorithm [44].

Interpolation of two-dimensional vector data using elasticity constraints is a method that allows you to create a smooth representation of sparse vector data. This method is based on the physical principles of elasticity, where the deformation of a material is determined by the forces acting on it. The method involves constructing a matrix of equations that relates observed data points to unknown data points, with the constraints of the elasticity equation. The matrix is then solved using a linear or non-linear optimisation algorithm to find the best estimate of the missing data points that satisfies the constraints. The elasticity constraints used in this method are usually derived from the Navier-Stokes equations, which describe the motion of the fluid and the forces acting on it. These equations are used to calculate a strain tensor that describes the deformation of the fluid at each grid point. The strain tensor is then used to calculate the stresses acting on the fluid, which are balanced by the forces acting on it. The method involves using elasticity constraints to interpolate vector data on a regular grid. This is done by fitting a smooth function to the data that satisfies the constraints. The smooth function is then used to estimate missing data points. The method uses a two-dimensional elastic model to provide a relationship between the two components of the horizontal velocity of GNSS points: latitude (N) and longitude (E). A vector force is assumed to act at each point. These forces cause deformation of an elastic body resulting in a vector deformation field. For any point in the field, the velocity vector is calculated, and the magnitude and direction of the acting force are adjusted until they are equal to a given value of the velocity vector. Coupled interpolation realises the response of an elastic body in two-dimensional space to an acting deformation based on Green's function analysis. Different states of elasticity (from elasticity to full compression) are adjusted using Poisson's ratio. The grid size is chosen so that each cell has at least one observation point [45]. The distance between points is 30-50 km, so a grid with a cell size of 1° was chosen to interpolate the velocity field components. The coefficient

Poisson's ratio was assumed to be 0.5, corresponding to a typical elasticity [46].

The results of the coupled interpolation method for building the horizontal velocity model of the region are presented in Fig. 1.8.

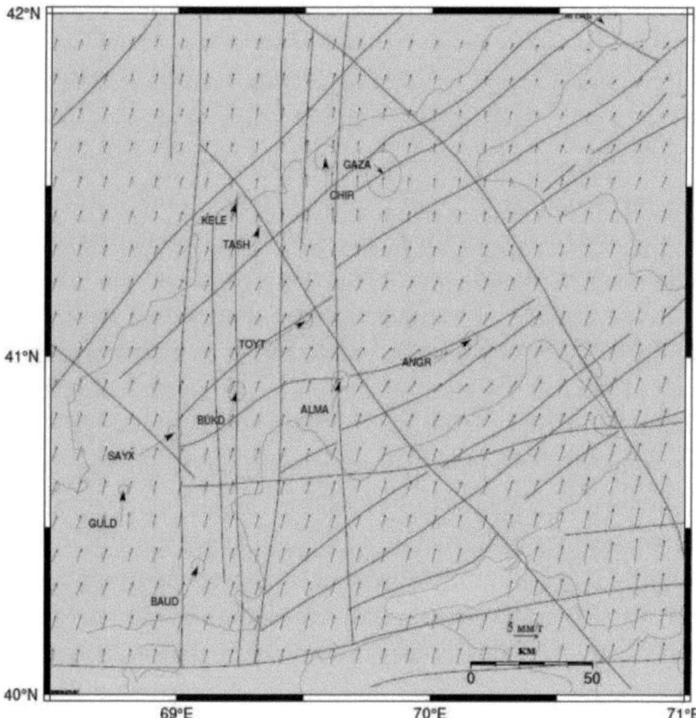

Fig. 1.8. Spatial distribution of the horizontal velocity field of the region (black arrows represent the original velocity field and blue arrows represent the results of interpolation)

Analysis of the spatial distribution of the measured velocity field of horizontal motion of the Earth's crust in the Tashkent region, despite the sparse density of points, revealed trends of non-uniform velocity distribution, which is confirmed by the geological structure of the region. In the zone of intersection of tectonic plates Karzhantau, Kumbelskaya and Chatkal, the tendency to rotational motion is seen. It is established that the range of values of horizontal displacements of points varies from 3 mm in the plain part to 10 mm in the mountainous part of the area.
Evaluation of the measured and interpolated numerical values of velocities confirms that the value of velocities in the east of the area is larger than in the western part. The general tendency of the direction of rotational motion of the investigated area is determined by clockwise rotation from west to east. The velocity field in the west increases from south to north. In the eastern velocity field, however, the numerical values take maximum values up to 10 mm/y. On the other hand, in terms of analysing the direction of motion, in the east, there is a constant motion from south to northeast, while in the west, the motion alternates with rotational blocks. In this paper, a study of the spatial field of GNSS point velocities of the Tashkent region based on observations from 2018 to 2020 is performed. The point velocities or the so-called velocity model obtained in the GAMIT/GLOBK programme allowed us to estimate the horizontal velocities of 14 points located in this region, the range of which varies from 21 mm/y to 33 mm/y. Relative to the "stable" Eurasian tectonic plate, the values of local displacements of the region were calculated, which can be both a consequence of microblock movement and the influence of anthropogenic factors (mining operations in the area of Angren and Almalyk). To obtain a continuous field of horizontal velocity distribution in the region, the method of coupled interpolation of two-dimensional vectors of the velocity field based on the Green's function and implemented in the cartographic programme GMT was used. A general trend of non-uniform velocity distribution was observed. Rotational movements along the Karzhantau, Kumbelskaya and Chatkal tectonic plates were revealed. The direction of motion tends to be clockwise from west to east. It was found that the range of values of horizontal displacements of points reached the minimum value of 3 mm in the plain part, and the maximum values up to 10 mm were observed in the mountainous areas of the region. The average velocity of

the stations in the region is about 4 mm/year. The methods and results of this study can be used both for the assessment of anthropogenic factors influencing the displacements of the earth surface during mining operations in the region and resulting from the high induced seismicity of the region, and for the implementation of the dynamic datum of the new national geocentric coordinate system of the republic [47-49].

Conclusions on the first chapter

The study analyses the spatial velocity field of 12 GNSS sites in the Tashkent region for 2018-2020. The horizontal velocity in the range of 21-33 mm/y was estimated. Local displacements related to microblock movement and technogenic factors (mining operations) were calculated. Using the coupled interpolation method, a continuous horizontal velocity field was constructed, revealing a non-uniform distribution. Rotational motions along several tectonic plates with a general clockwise trend were found. Horizontal displacements range from 3 mm in the flat part to 10 mm in the mountainous regions. The average station rate is approximately 4 mm/year. The results are applicable for the assessment of anthropogenic impacts and creation of a dynamic reference datum of the new national geocentric coordinate system.

CHAPTER II

VERTICAL ACCURACY OF DIGITAL ELEVATION MODEL (DEM)

2.1 Investigation of the influence of digital elevation model accuracy on tectonic processes

The land surface plays a fundamental role in modulating Earth's dynamical systems, including a large number of atmospheric, geological, geomorphic, hydrological and environmental processes. The extent and regulation of surface processes such as climate and tectonic forcing are constrained by the topography or shape of this surface [50; pp. 1173-1177]. Moreover, the strength of the relationship between shape and process can vary from weak to strong and is not always initially visible on the landscape. Consequently, understanding the nature of the land surface can provide insight into the nature and magnitude of the aforementioned processes [51; pp. 29-50].

Representing topography using digital elevation models (DEMs) is an important task for understanding tectonics, soil genesis and mapping, land cover, water flow, drainage and flooding, slope steepness, and soil erosion [52; pp. 1-336]. A DEM can be represented as a raster file in which the value of each pixel is associated with a certain topographic elevation. The main sources for DEM construction are contour lines, topographic maps, GNSS, photogrammetry methods, radar interferometry, and stereo images. These methods are compared on four aspects: cost, accuracy, sampling density, preprocessing requirements, and have both advantages and disadvantages. The main advantage of aerospace observations is the ability to obtain spatial, spectral and temporal resolution at relatively low cost.

A DEM is a three-dimensional computer graphic representation of elevation data for terrain representation [53; p. 775-780]. A global DEM refers to a discrete global grid. DEMs are often used in geographic information systems, and are the most common basis for digital elevation maps. The first DEMs used stereoscopic SPOT satellite images. In 2003, NASA released the SRTM v3 DEM dataset with near global coverage at a resolution of one arc second (approximately 30 m). The vertical error of the SRTM data is less than ±16 m [54; pp. 49-59]. The data are projected in geographic projection (latitude/longitude) with WGS84 horizontally and EGM96 vertically.

The ASTER GDEM V2 product is also a freely available DEM covering the land surface between 83°N and 83°S. ASTER GDEM is distributed in GeoTIFF format in geographic coordinate system (latitude/longitude) with a resolution of 1 arc second (approximately 30 metres). The data coordinate system is WGS84/EGM96.

Interferometric Synthetic Aperture Radar (InSAR) is another commonly used method for creating DEMs and topographic maps. InSAR uses two or more high spatial resolution SAR radar images to create high quality DEMs using phase interferometry techniques. InSAR imagery is accurate to within 1 metre. The survey is carried out regardless of weather conditions and at any time of day [55; pp. 857-872].

The scope of such data depends on the geographical extent of the coverage area, and before its use, the user should be aware of the impact of errors (incomplete observation density, positioning inaccuracy, data input errors, error handling, classification problems) of DEM in the study area [56; pp. 61-69]. Errors in DEM mainly consist of two components, horizontal, X and Y positioning accuracy and vertical errors or attribute data accuracy [57; pp. 578-595]. However, positioning accuracy and attribute accuracy cannot be separated, the error can be due to wrong height value for the correct position or correct height value for the wrong position or any combination of them. There are three main groups affecting the inaccuracy of SRTM and ASTER DEMs. The first group relates to the system parameters during data acquisition: baseline length and orientation, phase, slope and antenna position. The second group is related to the processing steps of the raw data, and the last group includes the influence of vegetation and land cover. For example, it is reported in [58; pp. 1378-1400] that an error of 1 angular second in baseline slope can be responsible for an error in height of 1.5 m at a distance of 300 km from the Earth. In addition, an error of 1 m in the length of the baseline will cause a 0.5 m height error.

Vertical height accuracy is the height difference between the modelled height and the actual height of the ground. Vertical accuracy depends on a wide range of terrain and land cover conditions, specific study areas, and the number of samples being evaluated. Most DEM assessments focus on absolute vertical accuracy rather than relative accuracy. Absolute vertical accuracy accounts for all effects of systematic and random errors and relates modelled height to true height in relation to an established vertical datum (georeferenced). Relative accuracy is the measurement of vertical accuracy within a particular dataset, i.e. the vertical difference between two points is measured and then compared to the difference in height for the same two points within the original dataset.

In altitude models, elevations are averages over a selected area. Thus, the spacing between points strongly affects the ability to maintain accuracy. For flat terrain, errors are minimal, but in hilly terrain, errors can be greatly increased. DEM products are constantly being improved and new versions are released. Usually, the newest versions are developed using analytical improvements that reduce the number of errors or inconsistencies found in previous DEM versions, especially based on evaluation studies [59, 60]. In recent years, many researchers have evaluated DEM products globally [61, 62; pp. 16-26].

Several processing algorithms and approaches have been proposed to combine with other elevation datasets to improve accuracy and eliminate systematic errors [64; pp. 57-67, 65; pp. 49-59, 66; pp. 5844-5853, 67; p. 2034]. However, although such products are widely used, they still show vertical errors that exceed acceptable errors. For example, modern global DEMs cannot detail elevation details when used for flood control [68; p. 1-169]. Kenward et al. (2000) evaluated the impact of different large-scale DEMs for hydrological runoff prediction and noted that different DEMs can result in a difference of almost 10% for runoff prediction [69; pp. 432-444]. Sanders (2007) investigated various publicly available DEMs for flood modelling, including US National Elevation Datasets (NED) data, SRTM, InSAR and LIDAR data. The author concluded that InSAR DEMs, including SRTM, are more prone to noise and require preprocessing, while NED DEMs are less prone to noise effects and can be used for flood estimation. Nevertheless, the author highlighted the usefulness of using SRTM as global terrain data sources for flood modelling [70; p. 18311843]. Another study evaluated the vertical accuracy of ASTER and SRTM DEMs and concluded that the delineation of the drainage network is significantly different compared to the real location [71; pp. 205-2017].

Elkhrachy I. (2018) used GNSS-derived reference points for the area of

Najran city, Saudi Arabia [72; pp. 1807-1817] to evaluate the vertical accuracy of SRTM and ASTER DEMs. The accuracy of the models were ±5.94 m and ±5.07 m for SRTM and ASTER, respectively. Further, the author
used a topographic map as reference points, the accuracy was ±6.87 m and ±7.97 m.

In Uzbekistan, the only available elevation data are topographic maps at different scales. For four regions (Kashkadarya, Chadak, Surkhandarya and Fergana Valley) a local digital elevation model (DEM) based on digitalisation of topographic maps at 1:500 000 scale and interpolation of elevation values has been developed. Nowadays, DEM plays a particularly important role in assessing the influence of surface processes on landforming in various human activity sites (geodynamic polygons, nuclear power plants, reservoirs, etc.) in the country [73]. In addition, this work compares the vertical accuracy of SRTM and ASTER digital elevation models for the construction of Kyzylsai and Tashtepa reservoirs, and recommends the use of SRTM DEM for mountainous regions and ASTER DEM for flat areas in the construction of reservoirs [73].

DEMs (such as SRTM and ASTER) are used as major sources of topographic information for many applications, including hydrological analysis and modelling, flood modelling and hazard mapping, geohazard analysis and landslide characterization. Despite this, the quality and accuracy of the DEMs used, and their suitability for these applications, have not been accurately assessed. The accuracy of DEMs can be assessed by comparison with ground control points that are measured using high accuracy equipment such as GNSS.

2.2 Constructing a correction surface to a geoid model EGM96

The aim of this part of the study was to quantify and compare the vertical accuracy of publicly available DEMs such SRTM30, ASTER GDEM2 and ALOS AW3D30 for the TGP area.

In addition to the DEM, the study used GNSS data from the Seismology Institute (SeisNet) and the State Grid (GGS (Figure 1.2). The dataset used to verify the accuracy of the DEM consists of 37 geodetic points collected during fieldwork in the region. The networks were measured using only the GPS system (Fig.2.1).

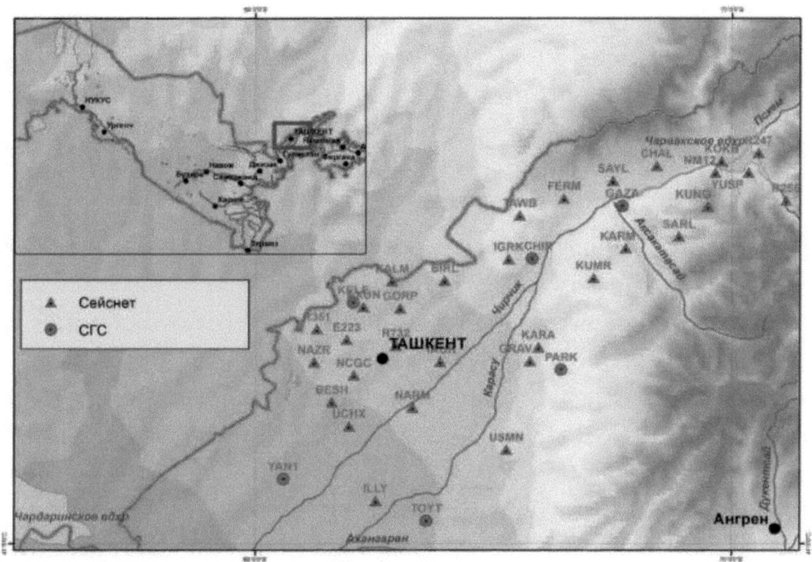

Seismnet stations of the Institute of Seismology of the Academy of Sciences of the Republic of Uzbekistan

GHS - Class 1 satellite network stations of the Republic of Uzbekistan

Figure 2.1. Network of GNSS geodetic points of the TGP

The GNSS network includes several subnets:

Seisnet: a special geodetic network of the Institute of Seismology in the TGP area, created to determine the distribution of the velocity and deformation field. Measurements were made by specialists of the National Centre of Geodesy and Cartography. Data from 31 GNSS stations collected during 6 cycles of measurements in the period from 2009 - 2010 were obtained. The average distance between the points is 5-30 km. Each measurement was made with the help of a receiver

Ashtech Z-Surveyor with simultaneous station observations for 20-30 days. The positions of the points for each epoch were estimated by the Institute of Seismology using Trimble Total Control (TTC) software version 2.73. Data from the permanent stations of the permanent IGS network (KIT3, MDVJ, POL2, NSSP) were used to relate the regional measurements to the global reference frame ITRF2014.

GGS: 6 GNSS/levelling points of the State Satellite Geodetic Network of 1st class points of the State Committee of the Republic of Uzbekistan on Land Resources, Geodesy, Cartography and State Cadastre. The GGS data were processed in the GipsyX/RTGx programme provided by Jet

Propulsion Laboratory using the high-precision point positioning (PPP) method [74; pp. 5005-5017]. The resulting daily point positions were georeferenced to ITRF2014. PPP with ambiguity resolution in both static and kinematic modes has an error with coordinate repeatability of 2 mm for the horizontal component and 6.5 mm for the vertical component [75; pp. 469-489].

In practice, normal heights can be determined by classical geometric levelling. The application of these methods is limited to distances due to the limitation of the observation range from the leveller, which varies only between 50 and 100 metres. Satellite levelling is used to measure distances between points from tens to hundreds of kilometres. The current level of development of GNSS technologies in Uzbekistan allows us to consider the possibility of replacing classical methods with high-precision satellite GNSS levelling and making measurements at short intervals. But, the ellipsoidal height determined by GNSS receivers can be converted to normal height only if the accurate geoid model of the given area is taken into account. And, despite the centimetre level of accuracy supported by GNSS receivers, in the absence of an accurate geoid model, the results of height network equalisation will be far underestimated. One of the methods used to construct the geoid has been to combine modern global models of the Earth's gravity field with GNSS measurements.

Molodensky M.S. proposed to measure normal heights relative to a quasigeoid - an equipotential surface coinciding with the mean level. Carrying out measurements with instruments, the vertical axis of which is directed along the plumb line, allows us to consider the geoid as a true figure of the Earth. But, for practical purposes, calculations are performed relative to another mathematical surface - an ellipsoid. A high-precision model of the geoid is necessary to determine the relationship of heights, gravitational potential, to establish a connection between local and global coordinate systems, to solve problems of geodynamics, etc. The geoid is a true physical surface of the Earth. The true physical surface of the Earth and the relationship between different surfaces is shown in Fig. 1.3 and is represented by the classical relation [33]:

$$h = H + \zeta \tag{2.1}$$

Fundamental equation (2.1) relates ellipsoidal heights measured by GNSS (h) to heights obtained in the local system by measuring gravity anomalies

(ζ) and levelling (H). The normal heights (H) must be calculated to an order of magnitude accuracy of the height determination by GNSS methods. This requires a highly accurate geoid model, which in turn can be divided into 3 components: long-wavelength, mid-wavelength parts and a correction due to topography. The first component is determined using global gravity models of the Earth, the second one using the Stokes integral and the third term using the DEM. The practical application of equation (2.1) is complicated by a number of factors, among which are random errors (noise) in determining the values of h, H and ζ , the difference of relativity surfaces for different types of heights and insufficient knowledge of the relationship between them, systematic distortions due to errors of the long-wave part of the geoid, unmodelled GNSS errors, errors of the equalisation network, geodynamic phenomena (i.e. land subsidence, deformation of tectonic plates near subduction zones, etc.).

Global geoid height model EGM96, derived from gravimetric measurements, available in the public domain on the website of the Calculation Service of the International Centre for Global Earth Models (ICGEM) [78; pp. 907-1205]. But its accuracy has not been assessed for the territory of Uzbekistan and, therefore, it may be insufficient for users' tasks. The lack of gravimetric study of the area further aggravates the problem of using global models without preliminary processing and evaluation. Polynomial model [79; pp. 47-72], GNSS/levelling [80; pp. 31-35] methods have been used in earlier studies to improve the efficiency of the global model. The method proposed in this paper uses a network of points with known elevations from GNSS/levelling and from EGM96 model data. The enhancement of EGM96 with GNSS data was performed by modelling and further interpolation to the whole domain of differences between GNSS and EGM96 heights at "common" points [81; pp. 85-91]:

$$\Delta N_i = N_i^{GPS-leveling} - N_i^{EGM96} \tag{2.2}$$

namely the construction of the so-called correction surface (correction

$$f(\varphi,\lambda) = \Delta N = N_i^{GPS-lev} - N_i^{EGM96} = a_0 + a_1\varphi_1 + a_2\lambda_i ... = a_i^T x + v_i \tag{2.3}$$

surface):

$$T_i = a_0 + a_1\varphi_1 + a_2\lambda_i + ... - \textit{trend surface} \tag{2.4}$$

Trend values were calculated for each point, then dN values were calculated by subtracting the difference in geoid heights:

$$dN_i = \Delta N_i - T_i \qquad (2.5)$$

As a result, the improved value of the geoid heights is equal:

$$N_i^{improv} = N_i^{EGM96} + dN_i \qquad (2.6)$$

were constructed using EGM96 1' x 1' global geoid height deviation grid (Fig. 2.2a) for the region (Fig. 2.26). The continuous surface was created using the Natural Neighbour interpolation method, which provides better geoid height accuracy in mountainous areas [28; pp. 29-33].

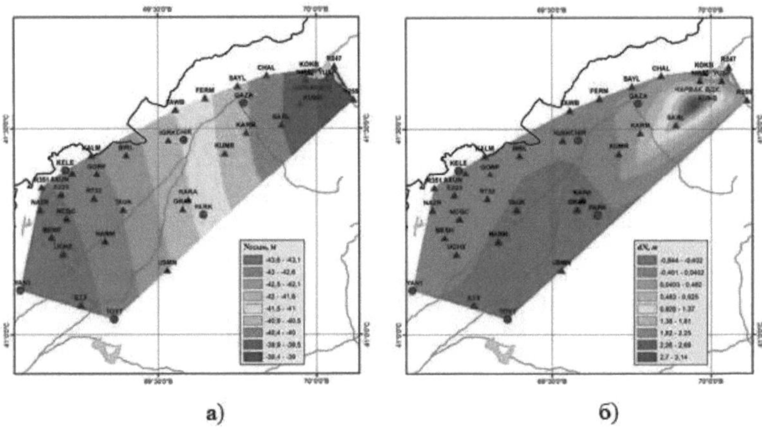

a) б)

Fig. 2.2. Values of N_{EGM96} geoid height deviations (a) and correction surface (*dN*) for the geoid model (b)

The range of geoid correction values varies from -0.86 m to 2.08 m with an average value of 0.84 m for the region. It should be noted that along the coastal zone of Charvak reservoir due to the lack of "common points" it was not possible to construct the correction surface.

2.3 Investigation of vertical accuracy of digital elevation models using GNSS

There are several types of DEMs that are used for different practical applications. A Digital Surface Model (DSM) can be useful for landscape modelling, urban modelling and visualisation applications. A digital terrain model (DTM) is applicable for flood or drainage modelling, land use studies [82; pp. 459-470], geological applications, and other applications. In most cases, the term digital surface model represents the earth's surface and includes all objects on it. Unlike a DEM, a DEM represents the bare surface of the earth without any objects such as plants and buildings [83; p. 210]. DEM is often used as a generic term for DEMs and DEMs that represent only elevation information without any additional surface definition. Most data providers (USGS, ERSDAC, CGIAR, SPOT Image) use the term DEM as a generic term for digital models. All datasets, whether from satellites, aircraft, or other flying platforms, are originally DEMs, such as SRTM or ASTER GDEM, although in forested areas SRTM penetrates the crown of trees, providing information somewhere between a DEM and a DEM.

The quality of the spatial resolution of a DEM is a measure of accuracy, how accurate the height of each pixel is (absolute accuracy) and how accurately the morphology is represented (relative accuracy). The following are some important factors that play an important role in the spatial resolution of a DEM:

1. Rough terrain
2. Sampling density
3. Grid resolution or pixel size
4. Interpolation algorithm
5. Vertical resolution
6. Terrain Analysis Algorithm

Currently, publicly available DEMs such as SRTM, ASTER, ALOS have almost global coverage in areas with insufficient observational data and in hard-to-reach areas. The scope of use of such data depends on the geographical extent of the coverage area, and before using it, the user should be aware of the impact of errors (incomplete observation density, positioning inaccuracy, data entry errors, error handling, classification problems) of the DEM in the study area [84; pp. 57-64].

The SRTM mission is a joint project of NASA and NGA. Interferometric radar image pairs acquired in 2000 by imaging from the reusable Shuttle spacecraft. The goal of this project was to acquire digital topographic data for 80% of the Earth's surface except for the northernmost (>60°), southernmost latitudes (>54°), with data points spaced at 1 arc second (approximately 30 m) on a latitude/longitude grid. The imagery was acquired with two SIR-C and X-SAR radar sensors. The absolute vertical accuracy of the data is 16 m.

A method called radar interferometry was used for the SRTM model. The method of satellite radar interferometry uses the effect of interference of electromagnetic waves and is based on the mathematical processing of several coherent amplitude and phase measurements of the same area of the Earth's surface with a shift in space of the receiving radar antenna. Differences between such images make it possible to calculate the height of the surface or its changes.

SRTM data are useful in applications where accurate surface shape and elevation are required. The data are also applicable in practical applications with flood prevention, reforestation, volcano monitoring, earthquake research and glacier movement monitoring.

ASTER GDEM, released in June 2009, was generated using stereo pair images collected by the ASTER instrument on board the Terra satellite. The ASTER GDEM coverage area extends from 83°S, covering 99 per cent of the Earth's land area. The vertical accuracy of the data is 15-20 metres. The enhanced ASTER GDEM V2 adds additional stereo pairs, improving coverage and reducing the appearance of artefacts. The enhanced product algorithm provides improved spatial resolution, increased horizontal to vertical accuracy. ASTER GDEM V2 supports the GeoTIFF format with 30 m resolution [85; p. 10]. ASTER GDEM2 stereo pair has a wider view at nadir when collecting data in steep and rugged terrain, which makes it better compared to SRTM30.

In May 2016. JAXA released the ALOS WORLD 3D (AW3D30) global digital surface model dataset with a horizontal resolution of approximately 30 m (1 arc second in latitude and longitude). Global Digital Elevation Model and Adjusted Rendered Images (ORI) created using archived data from the PRISM sensor. PRISM consists of three panchromatic radiometers that transmit stereo images. The spatial resolution of these images was 2.5 metres. JAXA calibrated the PRISM results with system adjustments to improve absolute accuracy and validate high-level products. Vertical accuracy averaged 7 m. ALOS AW3D30 dataset in the WGS84 ellipsoidal vertical coordinate system. This dataset was converted from the orthometric version using the available EGM96 geoid model. Table 2.1 summarises the general characteristics of some global DEMs.

Table 2.1.

Characteristics of global DEMs

№	Model	Developers	Data access	Ext.	Accuracy of models

				resolution (m)	
1	ALOS WORLD 3D 30	NTT DATA, RESTEC (Japan)	On a commercial basis	30	Vertical accuracy - 7 m Horizontal accuracy - 7 m
2	SRTM	NASA, NGA (USA)	Freely available	30	Vertical accuracy - 16 m Horizontal accuracy - 20 m
3	ASTER GDEM	METI (Japan), NASA (USA)	Freely available	30	Vertical accuracy - 15-20 metres Horizontal accuracy - 1520 m

The following publicly available DEMs are used in this paper to estimate the vertical accuracy:

SRTM30 v. 2.0 is a global digital elevation model (DEM) that includes a combination of data from the February 2000 Shuttle Radar Topography Mission and the GTOPO30 dataset from the U.S. Geological Survey. It can be thought of as the SRTM30 dataset enhanced with GTOPO30, or as an update of GTOPO30. It is a 1 arc-second (approximately 30 metres) resolution DEM provided in 1°x1° tiles by the National Aerospace Administration and the National Geospatial-Intelligence Agency (NGA) with the participation of the German and Italian space agencies. For each terrain segment, interferometric data were systematically collected at least twice from different angles (ascending and descending orbit) to fill in areas obscured by the radar beam shadow. The finished product, currently distributed by NASA/USGS, contains areas of "no data" (voids) where water or dense shadow interferes with altitude determination. Vertical accuracy averages better than 10 m (LE90). Two SRTM30 elevation data tiles for the study area were downloaded from the web site httpshttps://www.usgs.gov/centers/eroswww.usgs.gov/centers/eros.

ASTER GDEM v.2 (ASTER GDEM2): a stereoscopic product with a resolution of 1 arc second (30 m) was created from 1,880,306 Level 1A scenes collected between 1 March 2000 and 30 November 2013. ASTER GDEM2 was created by stacking all individual cloud-masked DEM scenes and DEM scenes without cloud masking, and then applying various algorithms to remove anomalous data. ASTER GDEM2 is distributed as

1°x1° tiles. Vertical accuracy is on average better than 15-20 m (LE90). The reason for considering this model is that ASTER GDEM2 has an advantage over SRTM30 because its stereo pair has a more nadir image when collecting data in very steep and mountainous terrain. For this study, 2 ASTER GDEM2 elevation data tiles were downloaded from the website httpshttps://asterweb.jpl.nasa.gov/gdem.aspasterweb.jpl.nasa.gov/gdem.asp .

ALOS World 3D (AW3D30): a global DEM with a horizontal resolution of approximately 30 metres (1 arc second latitude and longitude) was created from 5 metre resolution data and released in May 2016 by the Japanese Space Exploration Agency (JAXA). Information archived from the Prismatic Stereomapping Instrument (PRISM) was used to create the DEM and orthovertically rectified image (ORI). PRISM consisted of three panchromatic radiometers that captured stereo images along a track. It had a spatial resolution of 2.5 metres in the nadir radiometer and provided coverage of the entire Earth's surface, making it a suitable candidate for accurate global DEM and ORI. Over the past 10 years, JAXA has been calibrating the system, adjusting standard PRISM results with the ultimate goal of improving absolute accuracy and validating high-level products. Vertical accuracy averages better than 7 m (LE90). This is a non-standardised version of the ALOS AW3D30 dataset provided in the WGS84 vertical datum. This dataset was converted from the orthometric version using the available EGM96 geoid model. For this study, 2-tile data were downloaded from the website https: //www.www.eorc.jaxa.jp/AL.jaxa.jp/AL O S/en/aw3d30/index. htm.

Further data processing was performed using ArcGIS software version 10.6.1 and Surfer version 16. After completion of the preprocessing step (local geoid modelling, see 2.2), ellipsoidal height values were obtained for each digital elevation model. The statistics for the range of heights, mean and standard deviation for SRTM30, ASTER GDEM2, ALOS AW3D30 and GNSS heights are presented in Table 2.2.

Table 2.2.

Statistics of DEM elevation values for the study area

	Min, m	Max, m	Cf. value, m	RMS, m	GNSS and DEM correlation equation	Correlation coefficient. R^2
GNSS	314,07	1912,84	666,55	16,33	-	-

SRTM30	304,63	1909,45	660,52	16,35	y=1.0018x-7.2123	0,9998
ASTER GDEM2	308,54	1906,20	661,30	16,30	y=0.9968x-3.1469	0,9998
ALOS AW3D30	306,07	1903,45	660,07	16,27	y=0.9956x-3.5438	0,9998

The overall elevation values range from 304 m to 1909 m with a standard deviation of about 16.3 m. A linear correlation was also calculated between the GNSS-measured ellipsoidal height values and the values obtained from the DEM for each model. The correlation coefficient R^2 was found to be high - 0.99 in all cases, and showed no significant differences between the three DEM models and the GNSS height data.

The difference between the values of ellipsoidal heights calculated from DEM and GNSS (inconsistencies) will result in an error for each point and can be used for more detailed evaluation of the characteristics of digital elevation models. Figure 2.3 (a-c) shows the resulting surfaces constructed using the Natural Neighbour method for GNSS - SRTM30, GNSS-ASTER GDEM2 and GNSS-ALOS AW3D30 [28; pp. 29-33].

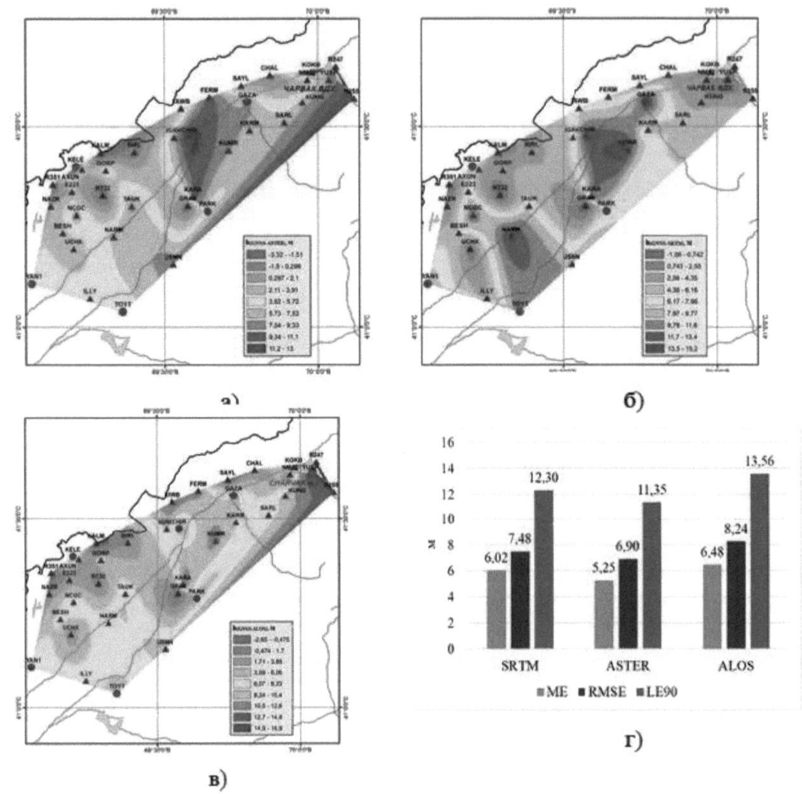

Figure 2.3. Difference between the distribution of ellipsoidal heights obtained by GNSS and DEM: a) SRTM30, b) ASTER GDEM2, c) ALOS AW3D30, d) DEM comparative statistics

The differences in elevation values were -1.06 m to 15.2 m, -3.32 m to 13 m, and -2.65 m to 16.9 m for SRTM30, ASTER GDEM2, and ALOS AW3D30, respectively. Note that the differences for all stations are within 3o (or RMSE). Absolute vertical accuracy with a probable linear error of 90% (LE90) was calculated from RMSE and ME. RMSE characterises the difference between the modelled Z_{MODEL} value and the reference values (in our case Z_{GNSS})- ME gives the difference between the simulated Z_{MODEL} value and the reference values.

Z_{GNSS})- ME gives an estimate of the bias from the reference model. The absolute vertical accuracy of LE90 can be estimated from the RMSE. The accuracy was estimated using equations [71; pp. 205-217]:

$$RMSE = \sqrt{\frac{1}{n}\sum_{i=1}^{n}(Z_{GPS} - Z_{MODEL})^2} \qquad (2.7)$$

$$ME = \frac{1}{n}\sum_{i=1}^{n}(Z_{GPS} - Z_{MODEL}) \qquad (2.8)$$

$$LE90 = RMSE * 1.6449 \qquad (2.9)$$

Accordingly, the best vertical accuracy of the three DEMs corresponds to their resolution. In the case of SRTM30 and ASTER GDEM2, the RMSE is comparatively higher than that of ALOS AW3D30. It is found that ASTER GDEM2 showed the lowest RMSE value of 6.90 m with a ME value of 5.25 m (Fig. 2.3d). It is found that SRTM30 and ASTER GDEM2 DEMs showed high absolute vertical accuracy of 12.30 m and 6.90 m, respectively. ASTER GDEM2 has high performance in changing statistical parameters compared to other DEMs. Although the displacement parameter ME is almost the same for SRTM30 and ASTER GDEM2 DEMs, according to the evaluation of the absolute vertical accuracy of ASTER GDEM2 DEM, it gives a more accurate estimate of the relief of the region (LE90 = 11.35 m). Therefore, it can be said that ASTER GDEM2 is the optimal DEM for the study area. Although all three used elevation models overestimated the true elevation marks, it is found that the mean and standard deviation in elevation difference of ALOS AW3D30 DEM is lower than that of SRTM30 and ASTER GDEM2 DEMs. Based on this, at this stage of vertical accuracy assessment, it can be tentatively concluded that SRTM30 and ASTER GDEM2 DEMs can be used for further deformation studies in this region.

To assess the accuracy of the local geoid model, the difference between ellipsoidal heights by DEM and GNSS for each point was analysed for separate parts of the TGP territory. The study area was conditionally divided into 4 main groups. The first part of the region is located in the coastal part of Charvak reservoir with nearby stations KOKB, NM12, YUSP, CHAL, KUNG, R247, R255, SAYL, FERM. The second and third parts of the region are separated from each other by the Chirchik River running between them and are hereinafter conventionally referred to as the

Northern Zone (IGRK, KALM, NAZR, NCGC, R351, R732, TAUK, TAWB, UCHX) and the Southern Zone (ILLY, GRAV, KARA, KARM, KUNG, KUMR, NARM, SARL, USMN). Finally, the last group of stations consisting of 6 GNSS/levelling points (CHIR, GAZA, KELE, TOYT, PARK, YAN1) of the GHS network (Fig.2.4).

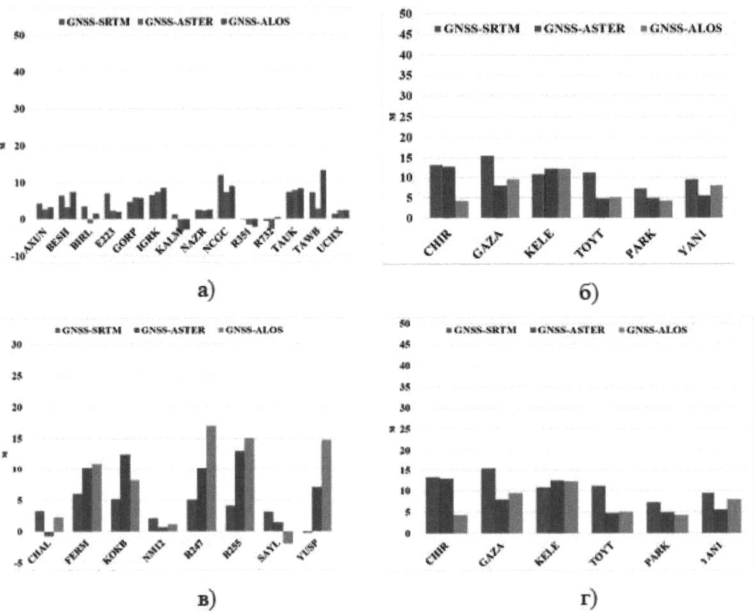

a) northern part; b) southern part; c) coastal zone of Charvak reservoir; d) points GNSS/levelling

Figure 2.4. DEM and GNSS height difference

For the coastal zone of the Charvak reservoir the maximum values of height difference equal to 13.01 m and 17.01 m were obtained for ASTER GDEM2 and ALOS AW3D30 DEMs, respectively (Fig. 2.4 c). This can be clearly seen at stations R247, R255, YUSP located in the eastern part of the reservoir.

The GHS network is mainly located in the central part of the study area, so the geoid surface needs to be refined in subsequent studies. In general, the results showed that the publicly available DEMs are quite satisfactory for geodetic and hydrological studies of the region. At the same time, it should be noted that the standard deviations of heights were worse for ALOS AW3D30 compared to SRTM30 and ASTER GDEM2. It is found that the accuracy of ASTER GDEM2 DEM is better than that of SRTM30 ALOS

AW3D30. ASTER GDEM2 DEM showed the lowest RMSE errors of 6.90 m with ME of 5.25 m. This is because ASTER GDEM2 has advantages over SRTM30 because its stereo pair has a large view at nadir when collecting data on hilly and rough terrain specific to our study area [86].

Conclusions to the second chapter

This chapter of the monograph presented the results of experimental research aimed at using GNSS measurements to estimate the present-day deformations of the Earth's surface and to improve the vertical accuracy assessment of the Digital Elevation Model (DEM). The region deformation rate and stress-strain state were estimated using the finite element method.

The EGM96 geoid model applied in the SRTM30, ASTER GDEM2 and ALOS AW3D30 DEMs was refined using "common points" of the satellite geodetic network. However, it is noted that the lack of data in the Charvak reservoir area affects the accuracy of the DEM in this area.

Analysis of the results showed that open source DEMs are suitable for geodetic and hydrological studies at TGP. ASTER GDEM2 showed better accuracy compared to SRTM30 and ALOS AW3D30, which is due to the wide view of the stereo pair on steep and rugged terrain typical of the TGP.

CHAPTER III

LINEAMENT ANALYSIS OF TECTONIC STRUCTURES

3.1 Introduction to lineament analysis and its role in tectonics research

According to the theory proposed back in the beginning of the last century by W. Hobbs, as a result of the stressed state of a lithospheric block, a structure of orthogonal fractures can arise, which manifest themselves as a network of straight or almost straight topographic elements of the regional extent of elements on the Earth's surface - lineaments [87; pp. 493-502]. Already the first Landsat satellite images back in the 70s of the last century allowed their mapping and laid the foundation of "fundamental lineament tectonics" [88; p. 16-17]. [88; pp. 16-17, 89; pp. 1-10]. Despite the fact that at first lineament analysis was used only for solving practical problems, such as the localisation of mineral deposits [90; p. 3041], a separate direction for the analysis of seismic tectonic processes began to develop at the same time [91; p. 749-767].

Determining where and how quickly deformation accumulates is a key requirement for improving the assessment of tectonic deformation and, in combination with geomorphological fault mapping, allows better identification of active structures. The key issues in lineament analysis are, firstly, to determine whether they actually form a globally uniform network of more or less orthogonal fractures and, secondly, whether they are tensile or compressional fractures.

As a rule, lineaments are linear or arc-shaped inhomogeneities of the Earth's crust and lithosphere of different rank, extent, depth and age of embedment, which are manifested on the Earth's surface directly (in the form of faults) or indirectly - by geological and landscape anomalies [92; p. 1-424]. A distinction is made between geological, geomorphological or topographical, artificial or non-geological lineament structures [92; p. 1-424, 93; p. 1-288]. Linear features of the earth surface caused by tectonic activity, representing faults, fractures or lithological boundaries are called geological lineaments. Topographic lineaments are lines caused by geomorphological processes such as drainage systems and mountain ranges. Pseudo or artificial features are roads, railways, crop field boundaries [94].

As a rule, lineaments on satellite images are the boundaries between different natural-territorial complexes, manifested as linear tonal (or coloured) anomalies with different width and severity. Moreover, complex lineament formations are strongly "blurred". There are also systems of short close subparallel lineaments and significantly wide (up to several hundred kilometres) lineament belts, which are interpreted on low spatial resolution images. Therefore, traditional methods of lineament structure extraction are labour-intensive and sometimes practically impossible due to both physical and geographical limitations. To date, remote sensing (RS) methods have proven their effectiveness in solving this problem and have found practical application in various fields [95; pp. 83-89, 96; pp. 137-142]. The degree of manifestation of lineaments in satellite images depends on the stress-strain state of the Earth's crust, which in turn determines the expression of structures in the physical and chemical properties of the Earth's surface. This is due to changes in surface temperature, soil moisture, and other properties of soil and rocks, as well as vegetation cover [97; p. 177-183]. To investigate active faults, the most reliable results are obtained by optical and radar measurements. The methods used to analyse lineament structures can be divided into the following: manual, semi-automatic and automated.

Manual or visual method is carried out based on the involvement of procedures to improve the image quality (directional filter, changing the contrast of the image, applying linear and nonlinear filters, using multispectral image features). For example, E.A.Ali et al. (2012) visually identified lineament structures from Landsat ETM+7 satellite data to recognise lithological and structural features in north central Sudan. Filtering and intensity hue saturation (IHS) transformation to improve spectral and spatial features provided regional lineament structures for geological map generation [95; pp. 83-89]. F. Al-Nahmi et al, (2016) performed lineament mapping using Landsat 8 images. Each spectral channel was assigned a specific target purpose, so for example, spectral channel 7 is better applicable for distinguishing geological deposits, spectral channel 5 for distinguishing soils from rocks and spectral channel 3 for distinguishing soil from vegetation. Noise reduction (MNF) was achieved using a combination of channels 4-6 and colour composition (RGB) of satellite image 7, 5, 3. Principal component method (PCA) was applied for the combination of channels 4, 5, 7 and band ratio 7/5, 6/4, 4/2. Directional filters were also used. This approach of satellite data processing allowed to explain the structural-geological and tectonic structure of the investigated area [96; p. 137-142].

The semi-automatic method of lineament structure extraction is based on image analysis using computer processing algorithms and visual interpretation. The processing steps consist of image visibility enhancement, classification based on reference objects, segmentation, Hough transform, PCA. Mallast U. et al, (2011) with a semi-automatic method for lineament extraction using linear filtering and object classification from DEM data, conducted a study of lineament outlines in the Black Sea area of Israel.

The authors used matrix filter and directional filter to improve the visibility of linear objects. They also used an object-oriented classifier with segmentation and vectorisation. And, in the final step, the automated Canny algorithm [98; pp. 2665-2678] was applied. Alshayef (2017) proposed the following procedure using Landsat-7 ETM+ images by semi-automatic method: inter-channel correlation (Decorrelation Stretching), Edge Enhancement, PCA. The results of the study confirmed the possibility of not only determining lineament structures, but also made it possible to assess the structural structure of the territory surface.

Hermi S.O. et al., (2017) proposed a method to extract lineament structures from Landsat 7 satellite data including 5 steps: image enhancement using histogram equalisation technique, use of Sobel filter, histogram segmentation, binary image generation and PCA analysis. The application of this method helped to identify several large faults in the study area, to determine the different morphological state of lineaments, and to identify the influence of tectonic movement on them [99; p. 440-455].

Automatic processing using special computer software systems includes image enhancement, filtering, edge detection and lineament extraction. Several algorithms for automatic extraction of lineament structures have been developed so far. These include Hough transform, LESSA, LEFA, TecLines, etc. The algorithms for automated extraction of lineament structures take into account noise, threshold, size and orientation of lineament structures. The results are presented in vector format and lineaments are now understood as linear zones of sharp changes in values (gradients) and axes of linear anomalies of the processed field, regardless of their origin. Formalisation of transformations (filtering, automatic lineament analysis, classification) and selection of lineaments is performed according to the features of the spectral brightness field. In particular, when analysing systems of spectral brightness gradients by "zones of correlation loss", lineaments transverse to the general gradient strike are clearly identified; gradient clumps of one direction may correspond to fault zones. The results of processing are presented in the form of maps in isolines, which facilitates their perception and further analysis.

One of the promising methods for monitoring tectonically active territories is the operational analysis of geodynamics by recording the variability of lineament systems identified in space images [100; p. 692]. In a series of published works, lineament analysis was used to study active fault patterns in hard-to-reach mountainous areas, to explore mineral deposits, and to analyse water resources [101; p. 197-206, 102; p. 44-48, 103]. Space methods are highly informative, in particular, when assessing the state of the geodynamic environment in the areas of artificial reservoirs [104; p. 75]. Areas with high lineament density are unsuitable for the construction of dams and reservoirs, since there is a high probability of water leakage, collapse of dam slopes, and the sedimentation rate will be higher [105; p. 81-103].

Earthquake sources are located in zones of intense and recent tectonic movements of the Earth's crust, and accurate mapping of geological features is an important task for understanding their mechanism. Earthquake sources are directly related to linear structural elements of the Earth's crust [106; p. 1-326]. Remote sensing techniques in recent years have shown great potential for mapping changes in geological features on a large scale [107; pp. 1501-1507, 108; pp. 476-480, 109; pp. 36-47]. The manual method of lineament extraction is a classical approach based on experience and knowledge, but it is inefficient in terms of time and labour. Automated and semi-automatic algorithms (segment tracking algorithm, Hough transform, PCI LINE, LEFA) have been developed to improve the efficiency. The semi-automatic and automatic approaches are based on linear and non-linear spatial filters, edge detection or on a recognition system [110; pp. 2665-2678]. Lineaments detected in automated remote sensing data processing have different physical nature. For example, more dynamic lineaments may reflect zones of compression and stretching of the Earth's crust [1131 pp. 47-56]. Analysis of statistical characteristics of lineaments (length, density and direction) is used to construct time series of lineament dynamics for seismic hazard monitoring [112; pp. 45-53]. The study of deformation waves at different time intervals (before, during and after the earthquake) has shown that the change of lineament structures starts 2-3 months before earthquakes, reaches a maximum about 20 days before the earthquake and ends 20 days after the earthquake. Moreover, lineament structures and their stress fields return to their original state 2-3 months after earthquakes [113; p. 3-20]. The most important factors affecting the earthquake level at the shock site are the magnitude, the distance to the earthquake epicentre and the influence of location [114; p. 677]. Many studies have focused on earth deformation studies using satellite data to determine the location and magnitude of seismic surface displacements, most of these earthquakes have been considered with magnitude Mw greater than or equal to 5.0 [115; pp. 1-26,116; pp. 1-14, 117; pp. 600-619, 118; pp. 1283-1291].

The main causes of earthquakes are natural processes (tectonic, volcanic) and artificial processes (mining operations, filling of reservoirs) [119; p. 54]. Depending on its size and location, an earthquake can cause physical phenomena in the form of shaking of the Earth, the appearance of faults on the Earth's surface, and in some coastal areas - tsunamis. The area within the Earth where faults form and seismic waves occur is called the earthquake source. The main seismic shock is preceded by preliminary weaker tremors called foreshocks. Smaller earthquakes, aftershocks, may follow the main shock, sometimes hours, months, or even years later [119; p. 60]. Four different types of waves propagate across and on the Earth's surface at different speeds, reach a location at different times, and cause vibrations in different ways. The first wave to reach the surface is the compressional wave or primary longitudinal seismic P-wave. The most destructive are transverse waves, S-waves, which propagate near the Earth's surface and cause the Earth to move at right angles to the direction of the wave and structures on its surface to oscillate from side to side. The third and fourth types of waves are slow, low-frequency surface waves, usually detected at large distances from the epicentre, that cause buildings to sway and water bodies to form waves [120; p. 8].

RS data started to be used for earthquake research from the 70s with the first satellite images. Mainly structural-geological and geomorphological studies were carried out. Maps of active faults and structures were produced. The disadvantage of the methods used was incomplete analysis of time series, which did not allow measuring short-term processes before and after earthquakes. Modern methods can analyse active faults using filtering. Earthquake studies using remote sensing indicate several phenomena related to earthquake processes, such as earth surface deformation, surface temperature and humidity, gas and aerosol content. Aftershocks record horizontal and vertical deformations ranging from tens of centimetres to metres. Such deformations are recorded by InSAR radar methods [121; pp. 38-44]. Pre-earthquake deformations are quite small, on the order of centimetres. There are numerous observations of surface and near-surface temperature, which increases by 3-5°C before earthquakes. Methods of earthquake prediction using thermal infrared (TIR) surveys are being developed. Satellite methods allow measuring concentrations of gases in the atmosphere: O_3 , CH_4 , CO_2 , CO, H_2S, SO_2 , HCl and aerosols [122; pp. 33-38].

Earthquake sources located in zones of recent tectonic movements of the Earth's crust and accurate mapping of geological features is an important task for understanding the mechanism of earthquakes. Special attention has been paid worldwide to the study of earthquake precursors. Developing methods for monitoring and predicting seismic hazard is crucial to prevent and reduce the number of deaths and damage caused by strong earthquakes. Many scientists have investigated the relationship of an impending earthquake to changes in the water table [123; pp. 21-60], changes in the acceleration of seismic waves by electromagnetic fields [124; pp. 1465-1468], movement of the earth's crust [125; pp. 115-121], release of radon and hydrogen [126; pp. 765-780, 127; pp. 61-63], large-scale changes in soil temperature [128], and changes in ion concentration in the ionosphere [129]. Some precursors can be recognised by the anomalous behaviour of various geophysical fields, including lineament dynamics [130; pp. 561-567, 131; pp. 295-328].

As a rule, lineaments form a well-defined network with several dominant directions. Lineaments are subdivided into transcontinental, interregional, regional and local.

Arellano et al. (2007) found the influence of earthquakes on the appearance of lineament systems. A significant number of lineaments appear almost one month before a large earthquake, and then the number of lineaments decreases one month after the earthquake. This feature was found in the analysis of 6 large earthquakes with magnitudes ranging from 5.2 to 7.8 in South America and China [132]. Singh V.P. and Singh R.P. (2005) used lineament analysis to study the change in stress structure around the epicentre of Bhuj earthquake (India) with Mw=7.6. The study used data from Indian LISS satellite (IRS-1D). Lineament structures were extracted from the images using filters. The results confirmed that the lineaments obtained during 22 days of the earthquake are different from those obtained 3 days after the earthquake. They are assumed to be associated with fractures and faults, and their orientation and density provide insights into rock features. The most important result is the high level of correlation between the continued maximum horizontal compressive stress obtained from lineaments and earthquake source mechanisms [133].

Image processing with extraction of lineament structures using various algorithms has been performed in [134; pp. 138-143, 135; pp. 5938-5958, 136; pp. 1121-1126]. The crustal dynamics tens of kilometres underground, reflected on the surface, have been investigated. Deeper phenomena extend over large areas and require images with low spatial resolution of 10-30 m for detection. These algorithms cannot isolate individual fractures, but they can integrate fault information over tens of kilometres and track changes due to the accumulation or weakening of force due to tectonic plate motion.

3.2 Lineament mapping and its structural interpretation: the example of the Charvak reservoir area

The study used optical images from the US Landsat 8 satellite, a collaboration between NASA and the USGS. Launched on 11 February 2013 from Vandenberg Air Force Base, California, on an Atlas-V 401, the satellite is in a sun-synchronous, circumpolar orbit with an altitude of 705 km, an inclination angle of 98.20° and a rotation period of 99 minutes. Equipment on board includes the OLI and TIRS instruments. OLI collects data in the visible, infrared and shortwave infrared spectral bands with spatial resolutions of 30 and 15 metres. TIRS collects data in the thermal spectral region with a resolution of 100 m. (Table 3.1).

Table 3.1.

Characteristics of Landsat 8 spectral channels

Spectral channel number	Wavelength (μm)	Spatial resolution(m)	Survey coverage range (km)	Temporary authorisation of one territory (day)
Channel 1 - Coastal / Aerosol	0,43-0,45	30	185	16
Channel 2 - Blue	0,45-0,51	30		
Channel 3 - Green	0,53-0,59	30		
Channel 4 - Red	0,64-0,67	30		
Channel 5 - NIR (near infrared)	0,85-0,88	30		
Channel 6 - SWIR 1 (shortwave near infrared)	1,57-1,65	30		
Channel 7 - SWIR 2 (shortwave near infrared)	2,11-2,29	30		
Channel 8 - PAN (panchromatic)	0,50-0,68	15		
Channel 9 - Cirrus SWIR.	1,36-1,38	30		
Channel 10 - Long Wavelength Infrared, TIR1 (far infrared)	10,60-11,19	100		
Channel 11- Long Wavelength Infrared, TIR2 (far infrared)	11,5-12,51	100		

Landsat 8 data are available on the USGS website. Image processing was performed using ENVI, PCI Geomatica, LEFA, ArcGIS, RockWorks. ENVI, developed by ITT Visual Information Solutions, is a leading geospatial image processing solution with advanced spectral tools, geometric correction, and GIS analysis. Written in IDL, ENVI provides flexibility in image processing [137]. PCI Geomatica is a software for remote sensing and photogrammetry data processing with a focus on fast processing [138; pp. 1-169]. LEFA in MATLAB performs lineament analysis and interpretation of tectonic faults. ArcGIS (ESRI) is used in geographic information tasks [139]. RockWorks is a tool for mapping, section modelling, data analysis, and more [140].

This chapter highlights the methodology and results of analyses of lineament structures in the vicinity of the Charvak reservoir. The choice of this area for deformation analysis is due to tectonic movements and water level fluctuations. The selection of Landsat 8 images is based on the states of the reservoir in March, June, September and December, covering the moments of maximum, minimum and gradual decrease of water level. An automated lineament analysis methodology using ENVI, PCI Geomatica and ArcGIS software is used to process the data. The process begins with the selection of raw data for the extraction of lineament structures, involving the use of multispectral imagery, contrast stretching and image enhancement. Special attention is given to the principal component analysis (PCA) method implemented using the ENVI software. PCA enhances a multispectral image by reducing data redundancy and modifying spectral dimensions. This method generates principal components, providing new spectral information, which significantly improves the interpretability of the data in the analysis of lineaments and deformation in the area [141; p. 581]. To achieve clarity in the representation of the land surface, it is critical to consider the standard deviation parameter. This parameter reflects the degree of deviation of object attribute values from their average value. Setting a certain value of the standard deviation allows to highlight those values that are above or below the mean, which in turn provides an effective and clear representation of objects on the surface. This approach makes it possible to highlight key features and anomalies, which is an important aspect of geo-information analysis and visualisation of Earth surface data.

In this study, the principal component analysis (PCA) technique was applied to seven spectral channels of Landsat 8 satellite (channels 1, 2, 3, 4, 5, 7 and 8), while the thermal bands (channel 6) were excluded from the analysis. The obtained PCA results are visualised in Figure 3.1. The component analysis performed indicated that the components of channels six, five and three were the most informative in terms of identifying geological structures, with a standard deviation equal to 1.5. This method allows the key aspects of the data to be highlighted, revealing values significant for geological investigations. Visualisation of principal components at a given standard deviation contributes to a better understanding of the geological structure and dynamics in the study area. High deviation component extraction allows the identification of anomalies and features, which is important for identifying potential geological structures and areas of interest. The results obtained open new perspectives for detailed analysis of the geological characteristics of the study area, highlighting the importance of the PCA method in geological research using satellite data.

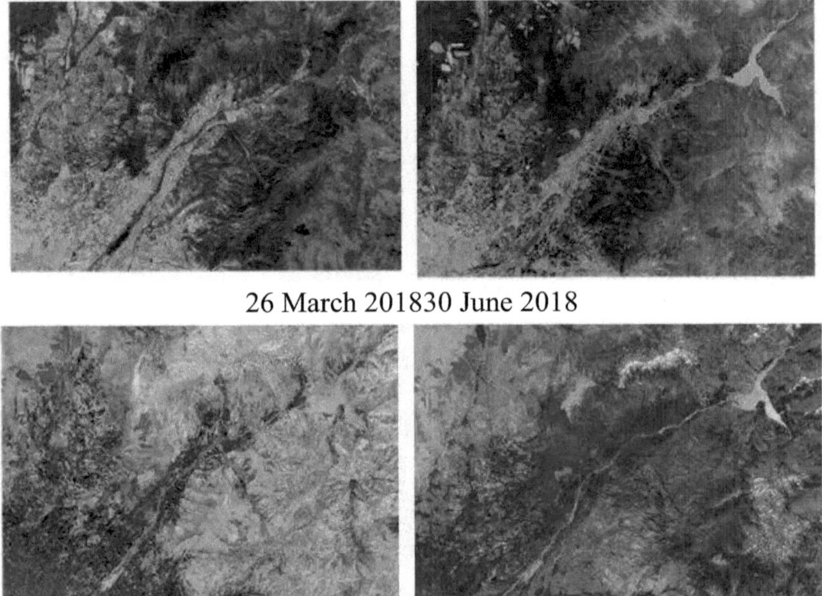

26 March 201830 June 2018

08 September 201813 December 2018

Figure 3.1. PCA results

The LINE module in the PCI Geomatica software [138; p. 84] was used to analyse lineament structures. The number and length of the extracted lineament structures depend on the values of the input parameters set in the LINE module. The characteristics of the six parameters of the LINE algorithm are presented in Table 3.1. After the lineament structures extraction step, statistical analysis, construction of rose diagrams and creation of density maps were performed.

The statistical analysis includes the evaluation of different characteristics of lineaments, such as their number and length, which contributes to a better understanding of the spatial distribution of structures. The pink diagrams are visualisations of the directionality of the lineaments, which helps to highlight the dominant trends in the distribution of linear structures (Table 3.2).

Table 3.2.

Parameter values of the LINE algorithm

№	Parameters	Unit of measurement	Value range	Values
1	RADI	Pixels	0 - 8192	10
2	GTHR	-	0 - 255	50
3	FTHR	Pixels	0 - 8192	30
4	LTHR	Pixels	0 - 8192	3
5	ATHR	Degrees	0 - 90	15
6	DTHR	Pixels	0 - 8192	20

A lineament density map created using spatial analysis tools in ArcGIS software provides information on the maximum accumulation of linear structures per unit area [142; p. 1-7]. This method of analysis complements the overall picture of the spatial distribution of linear structures by providing additional information on their density in the study area. Lineaments extracted from satellite images reflect both geological lineaments (faults and fractures) and hydrological structures (rivers or shorelines) [143; p. 24]. The results of automated extraction of lineament structures from PCA images are presented in Figure 3.2.

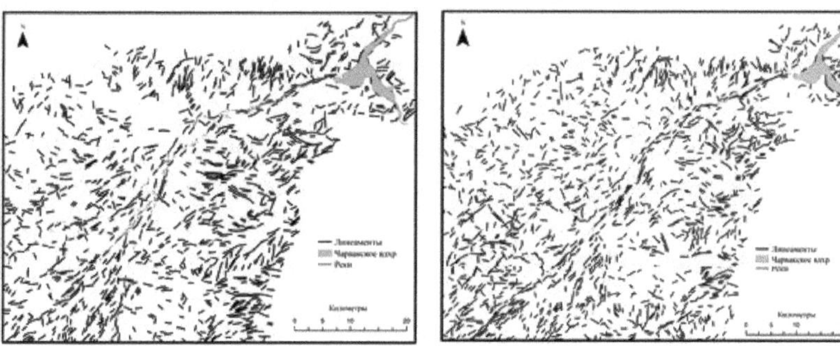

26 March 201830 June 2018

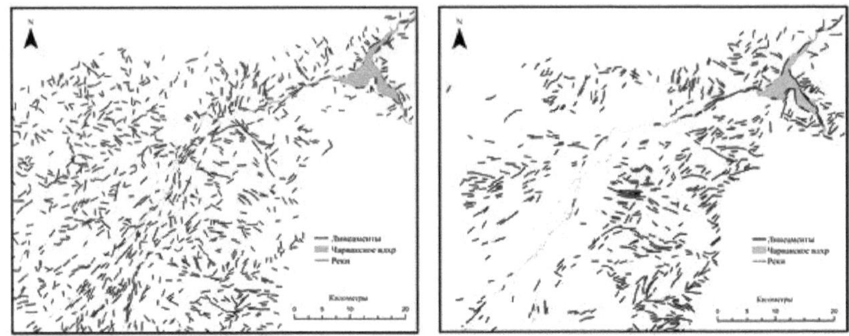

08 September 201813 December 2018

Figure 3.2. Extracted lineament structures from PCA images

The statistical analysis of lineament structures for the selected time period is presented in Table 3.3.

Table 3.3.

Significance of statistical analysis of lineament structures by histogram analysis

	December	March	June	September
Total number of	2250	4946	4821	4977
Min. length, m	30	42,4	30	60
Max. length, m	4031,2	3660,7	3606,2	4494,9
Amount, m	1952567,8	3919342,5	3694166,2	3665247,5
Standard deviation, m	476,5	426,6	428,3	406,6

It was revealed that in December, during the water level decrease in the Charvak reservoir, the number of lineaments was minimal and amounted to 2250. In March, June and September, a high number of lineament structures was observed. Probably, water level fluctuations have a greater influence on the formation of lineament structures than the periods of reservoir filling and discharge. Analyses of lineament structures at different time periods revealed their main direction, as presented in Figure 3.3.

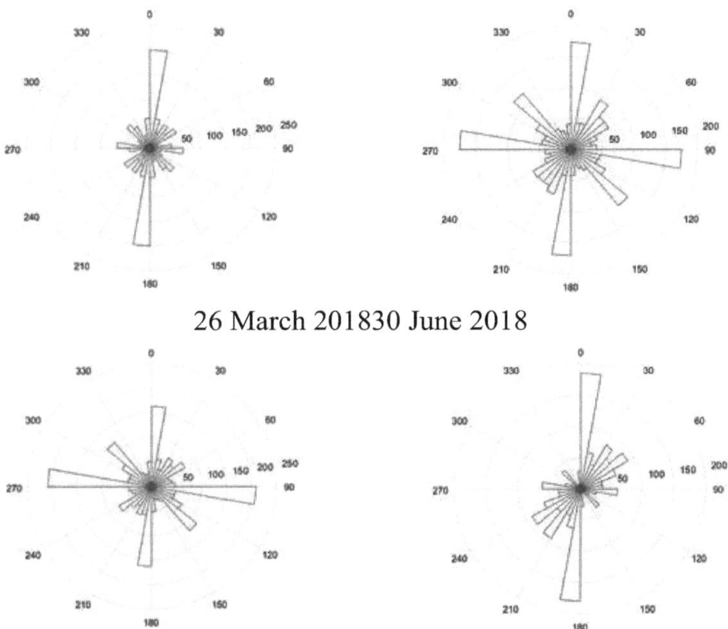

26 March 201830 June 2018

08 September 201813 December 2018

Figure 3.3. Rose diagrams of lineament structures

The results of interpretation of lineament structures showed that the predominant direction throughout all time periods was north-south (NS) direction. In addition to the main direction, a secondary west-east (WE) orientated direction was also identified throughout the time intervals. The north-south direction of the lineament structures indicates the movement of faults in the region. It should be noted that the west-east direction is observed during periods of high water levels in the Charvak reservoir (June, September), and this direction may indicate changes in water levels. Lineament density maps of the lineament structures of the study area show the lineament concentrations of the Charvak reservoir in more detail (Fig. 3.4).

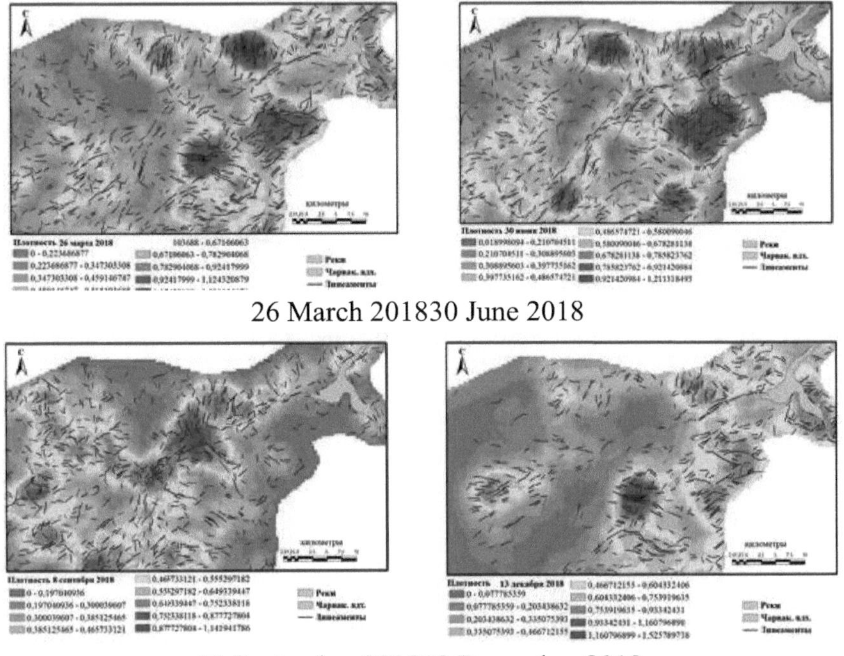

26 March 201830 June 2018

08 September 201813 December 2018

Figure 3.4. Density maps of lineament structures

The obtained density maps of lineament structures show predominantly increased density in the southeastern part of the study area. In the northern part of the region, the high concentration of lineament structures in the winter-spring period appears to be the result of abundant seasonal processes such as snow cover and precipitation. The density maps of lineament structures for June and September can clearly reflect faults resulting from water level fluctuations in the Charvak reservoir [144].

Conclusions to the third chapter

This chapter of the monograph presents in detail the results of experimental research on an automated method for the extraction of lineament structures in the TGP area. The developed method is based on a multilevel approach that includes the following image processing stages: image construction by principal component analysis and independent component extraction, as well as extraction of lineament structures using the LINE algorithm.

The results of seasonal analysis of lineament structures allow us to make an important observation: the minimum number of lineaments was recorded in December (2250) during the period of water release from the Charvak reservoir. In March, June and September a significant increase in the number of lineament structures is noted. It is concluded that water level fluctuations have more significant influence on lineament formation than periods of reservoir filling and release.

Density maps of lineament structures visualise predominantly increased concentration in the south-eastern part of the study area. The northern part of the region shows a high density of lineaments in the winter-spring period, which is probably related to intensive seasonal processes such as snow cover and precipitation. Lineament density maps for June and September vividly reflect faults resulting from water level fluctuations in the Charvak reservoir.

CHAPTER IV

LINEAMENTS AS ONE OF THE PRECURSORS OF EARTHQUAKES

4.1 Lineament structures as earthquake precursors: example of the Tashkent geodynamic polygon

The purpose of this study includes several key aspects: firstly, to create a regional structural map of lineaments in the study area using Landsat 8 optical images; secondly, to analyse the distribution of changes in lineament structures using rose diagrams and lineament density maps. This will make it possible to identify the precursors of seismic activity in the territory of the Tashkent geodynamic polygon (TGP) in Uzbekistan. Figure 4.1 shows that seismicity in this region increases from north to south, predominantly along fault lines such as the Karzhantau and Chiltinsky.

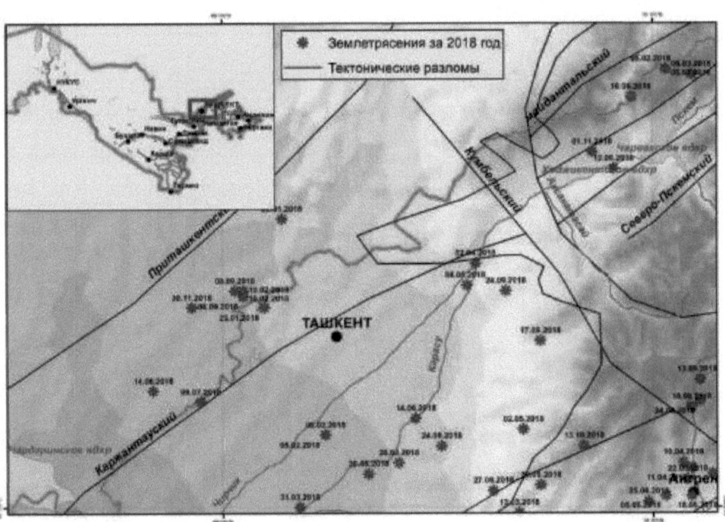

Fig. 4.1. Tectonic faults and epicentres of earthquakes in the region researches

The analysis was based on the catalogue of earthquakes from the Republican Seismoprognostic Monitoring Centre of the Ministry of Emergency Situations of the Republic of Uzbekistan [145]. When selecting earthquakes for analysis, the following criteria were followed. Firstly, events with a magnitude greater than 3 that occurred exclusively within the study area were taken into account. Secondly, dates with no satellite images or significant cloud coverage were excluded. The results of five earthquakes in bold were selected from the catalogue (see Table 4.1).

Table 4.1.

Catalogue of earthquakes recorded in 2018 in the territory of TGP

№	Date	Time	Depth (km)	Magnitude, Mw	Location of the epicentre
1	06.01.2018	00:24:12	15	3	Kazakhstan
2	08.01.2018	22:26:00	25	3.5	Central Uzbekistan
3	08.01.2018	23:02:00	25	4.1	Central Uzbekistan
4	14.01.2018	03:24:00	10	3.3	Central Uzbekistan
5	23.01.2018	01:26:08	9	4.3	Central Uzbekistan
6	05.02.2018	19:53:45	4	3.4	Central Uzbekistan
7	10.02.2018	20:18:47	12	3.1	Central Uzbekistan
8	20.02.2018	13:18:25	6	3.3	Central Uzbekistan
9	24.02.2018	11:41:39.3	24	3.7	Central Uzbekistan
10	10.03.2018	09:32:10	5	3.1	Central Uzbekistan
11	13.03.2018	04:46:43	9	3.3	Central Uzbekistan
12	25.03.2018	12:51:20.6	9	3.3	Central Uzbekistan
13	05.04.2018	10:26:00	15	3.6	Central Uzbekistan
14	**02.05.2018**	**01:19:25**	**5**	**3**	**Central Uzbekistan**
15	**10.05.2018**	**10:56:11.5**	**29**	**3.1**	**Central Uzbekistan**
16	**22.05.2018**	**09:47:20.5**	**7**	**3**	**Central Uzbekistan**
17	12.06.2018	13:53:08	15	3	Central Uzbekistan
18	25.06.2018	14:12:08	10	3.4	Central Uzbekistan
19	**27.08.2018**	**13:22:07**	**5**	**3.1**	Central Uzbekistan
20	**08.09.2018**	**20:39:56**	**5**	**3.1**	**Kazakhstan**
21	30.11.2018	13:16:48	13	3	Kazakhstan
22	19.12.2018	10:33:59.2	7	3.4	Central Uzbekistan

The current study used satellite imagery from Landsat 8, the data of which was obtained from the Earth Explorer website of the United States Geological Survey (USGS). Table 4.2 presents the dates of Landsat 8 imagery acquisition and the number of days before and after the recorded events. It should be noted that satellite images for 2 May, 10 May, 22 May, and 27 August are missing. Therefore, the closest dates for interpretation were selected, namely images from 3 May and 23 August.

Table 4.2.

Dates of Landsat 8 satellite images used

Date of images	Period	Number of days (before and after the earthquake)
03 May.	Before and after the earthquake	1 day after and 7-19 days after
07 Aug.	Before the earthquake.	20 days before the earthquake
23 August	Before the earthquake.	4 days before the earthquake
27 August	Earthquake	No image
08 September.	Earthquake	-
24 September.	After the earthquake.	16 days after the earthquake
26 October	After the earthquake.	48 days after the earthquake (~1.5 months)
11 November	After the earthquake.	64 days after the earthquake (~ 2 months)

The RED, NIR, and SWIR 1 spectral channels with 30 m spatial resolution were combined with a panchromatic channel of higher spatial resolution of 15 m using the Create Pan tool in ArcGIS software. This improved the spatial characteristics of the lower spatial resolution channels. To create a regional map of lineament structures in the study area, an automated method for identifying lineament structures was used [130; pp. 561-567]. This method is based on the interpretation of geological structures that appear on the surface in the form of circular and linear structures of different degrees of curvature. The open-source LEFA software was used for image processing. The software includes many image processing steps starting from pre-processing. It applies algorithms to identify contours, extract linear elements, consider their number, and combine collinear linear elements into lineaments. The programme also includes the step of finding the fractal dimension of the image contours. At the final stage, the map image is exported in (.geotiff) format to linear elements in (.shp) format. This integrated approach provides a complete cycle of image processing in order to extract and analyse linear structures in the image [146, 147].

The algorithm includes several steps [146; pp. 8-22]:

1) *Image filtering:* filtering, including the use of graphic filters, was applied to prepare the image for edge detection calculations. For Landsat

images in shades of grey, sharp linear changes in brightness were identified and interpreted as linear structures. A highly efficient Canny algorithm was applied to identify these structures. An optimal filter using the sum of four exponents was chosen to isolate, localise and minimise repetitive responses at a single edge. Pixel boundaries where a local maximum of the gradient in the direction of the gradient vector is reached were isolated. This process involved the use of a filter based on Gaussian first derivative [148; pp. 679-698]. The Canny detector used a filter that provides effective noise suppression and meaningful boundary extraction. The points where a local maximum of the gradient is reached are considered as pixel boundaries. Suppression without a maximum helps to highlight the most significant boundaries, which is particularly important when analysing satellite images. Thus, the entire filtering and edge detection process has been carefully tuned using optimal parameters and techniques to ensure the accuracy and efficiency of line structure extraction in Landsat 8 images.

2) The probabilistic Hough transform (PHT) method was used *to digitise the linear feature vector* as described in [137; pp. 11-15]. This method provides an efficient way to extract linear structures in images. The parameters defined for the Hough method were carefully chosen to achieve optimal results:

- Minimum length of detected lines (in pixels): defines the minimum length of line structures to be selected in the image. It is set to a specific value to exclude short and irrelevant lines.
- Maximum pixel gap tolerance for a particular line: adjusts how much pixel gap tolerance is allowed for line detection. It helps to account for possible discontinuities or brightness changes on line structures.
- Select Hough coordinate values (p and 0): defines the range of p and 0 values for searching in Hough space. Effective selection of this range is important for accurate line detection.
- Number of Hough peaks ("Hough peaks"): determines how many peaks in Hough space are considered sufficient to detect a line. This controls the sensitivity of the method.

3) Large linear elements - faults - are obtained by combining several elements. The merging parameters are spatial proximity of these segments and their collinearity. The collinearity algorithm is applied to linear elements with length more than several pixels and is controlled by the

following parameters: the maximum difference k/b (the ratio of the slope of the equation of the compared lines to the intersection point), the maximum distance between the centre points of the compared lines, the minimum number of lines needed to merge into lineaments, and the order of the line polynomial.

This method of analysing satellite imagery allows the creation of a regional structural map of lineaments, which is an important step for the analysis of geological features and stress-strain state

The study area is a tectonically active zone with frequent earthquakes with an average magnitude of 3. Table 4.3 presents the general statistical information of lineament maps for all time images (7 scenes). The results of the statistical data analyses presented in Fig. 4.2 emphasise the influence of earthquakes on the linear structures in the study area.

Based on the analysis of the statistics for 3 May 2018 in the context of the three earthquakes that occurred on 2 May 2018, 10 May 2018 and 22 May 2018, it can be observed that the number of lineaments was 1352. This result shows that in the 20 days before the earthquake on 27 August 2018 (7 August 2018), the number of lineaments decreased to 97. In the 16 days before the earthquake, 23 August 2018, the number of lineaments increased to 571. On the day of the second earthquake, 8 September 2018, 1,494 lineaments were identified, and 16 days later (24 September 2018) the number decreased to 117. A noticeable decrease in the number of lineaments is observed after the October and November earthquakes, reaching 85 and 49 respectively.

It is important to note that the values of the minimum and maximum lineament lengths remain almost constant for all the events considered. These statistics emphasise the influence of earthquakes on linear structures in the region and provide information on the dynamics of changes in the tectonically active zone.

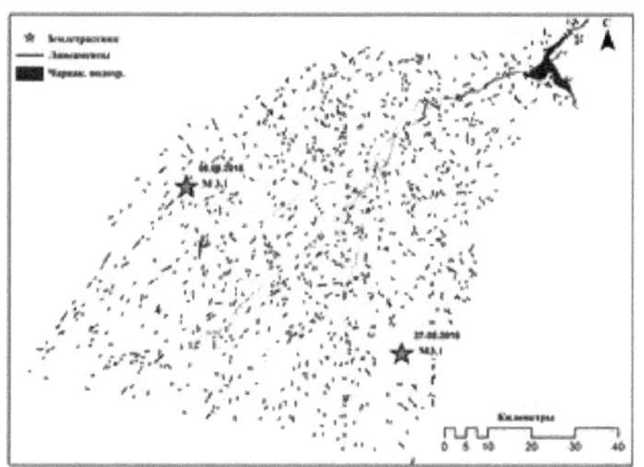

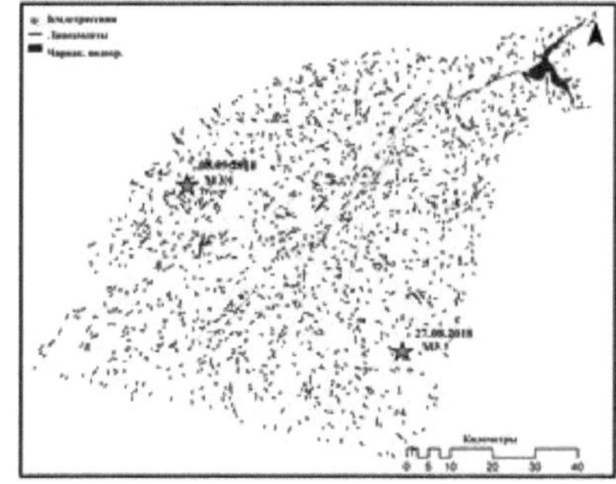

07 August 201823 August 2018

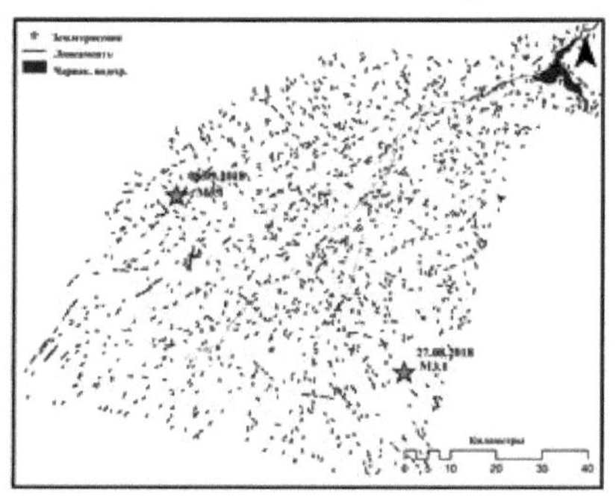

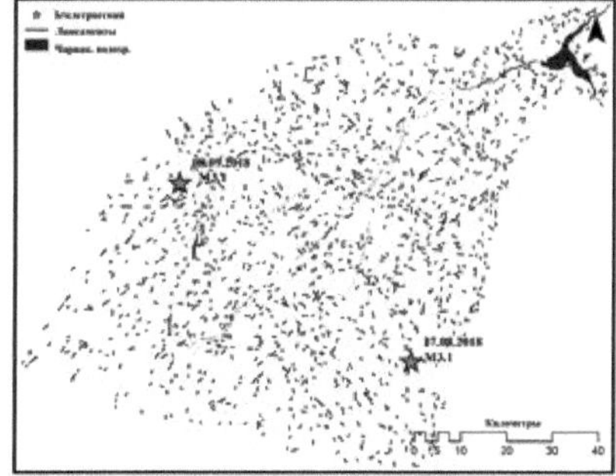

08 September 201824 September 2018

Figure 4.2. Result of lineament structures interpretation

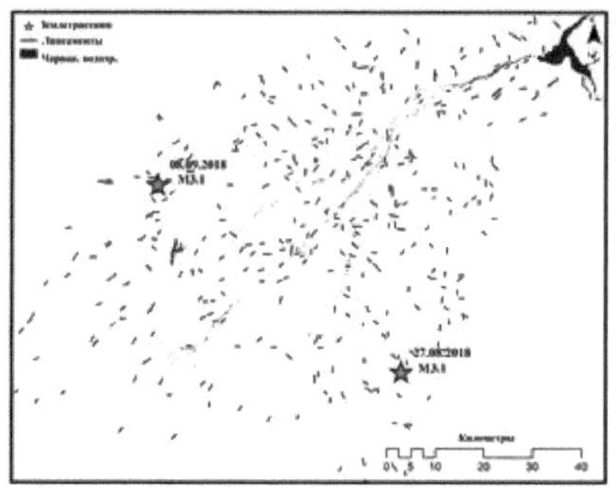

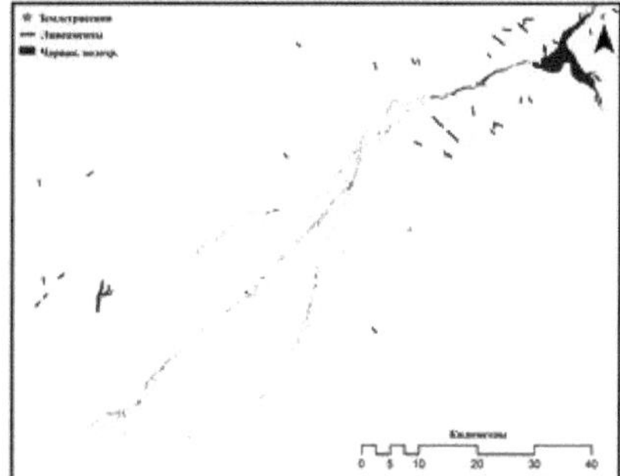

26 October 201811 November 2018

Figure 4.2. Lineament structures interpretation result (continued)

Table 4.3.

Lineament length statistics

	03 May.	07 Aug	23 Aug.	08 sept	24 Sep.	26 Oct.	11 Nov.

Total number of	1352	97	571	1494	117	85	49
Min length, km	0.8	1.5	1.1	0.8	1.5	1.5	1.1
Max length, km	4.1	3.0	3.6	3.8	3.6	3.0	2.8
Amount, km	1333.8	177.6	748.9	1457.6	204.9	153.9	66.1
Standard deviation, km	0.3	0.4	0.3	0.3	0.3	0.3	0.4

Obviously, the increase in the density of lineament structures is associated with tectonic fractures. Analysis of the statistics of the number of lineament structures shows that their number starts to increase almost 20 days before the earthquake, reaches a maximum about 4 days before the event and decreases 16 days after the earthquake, with the minimum value observed 2-3 months later.

The density of lineament structures depends on their number, making rose diagrams and density maps key tools for predicting change. Figure 4.3 explores the dominant directions identified from the data. The NNW, WNW, NW-SE and NE-SW directions are marked. The distribution of lineaments in these directions is related to the geological features of the region.

The SJ direction related to the Eurasian plate motion becomes noticeable 20 days before the earthquake, persists during the events and practically disappears after 2 months. Changes in the SW direction are manifested with the same periodicity, but with different intensity, especially in summer period, which is probably related to the processes in Charvak reservoir. NW-SE and NE-SW faults are caused by tectonic disturbances of mountain massifs. The Talaso-Fergana fault, the Kumbel fault, and other systems play an important role in the deformation and rotation of mountain systems [149; p. 54]. These observations emphasise that changes in lineament structures reflect complex processes in the Earth's crust, including tectonic movements and the impact of human activities, such as the impact of reservoirs [150].

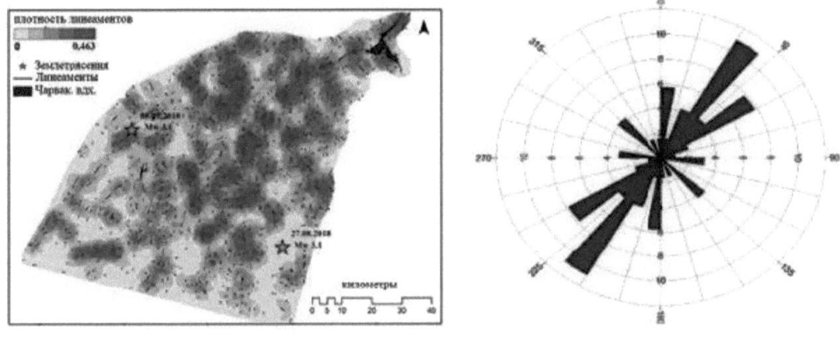

07 August 2018 07 August 2018

Figure 4.3. Density maps and rose diagrams of lineament structures

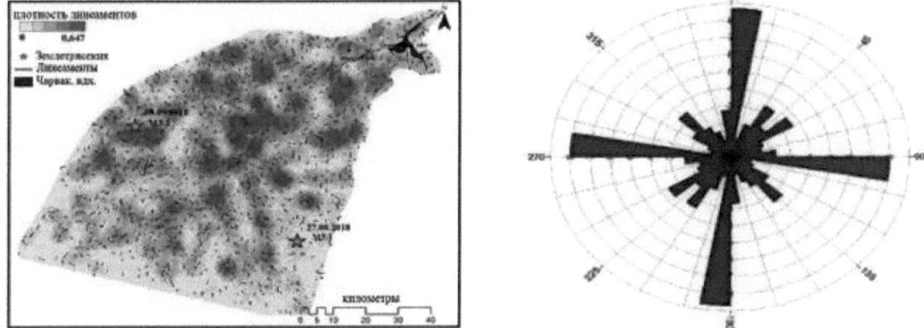

23 August 2018 23 August 2018 23 August 2018

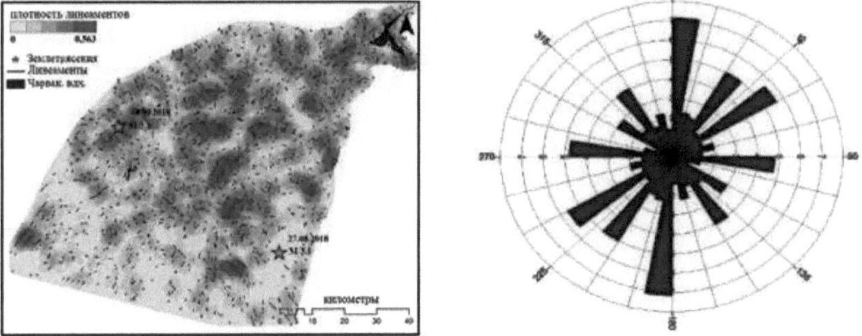

08 September 2018 08 September 2018

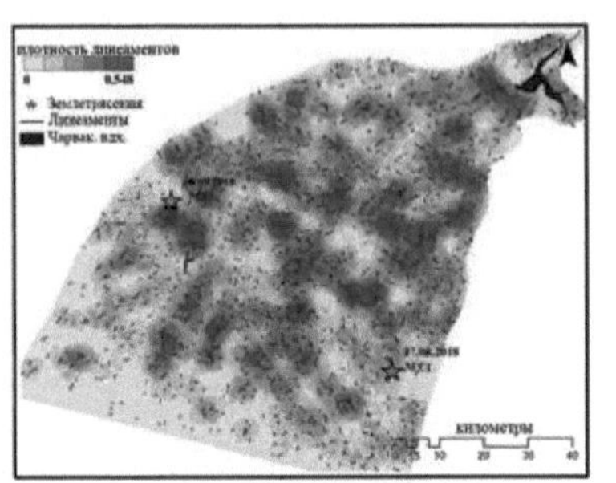

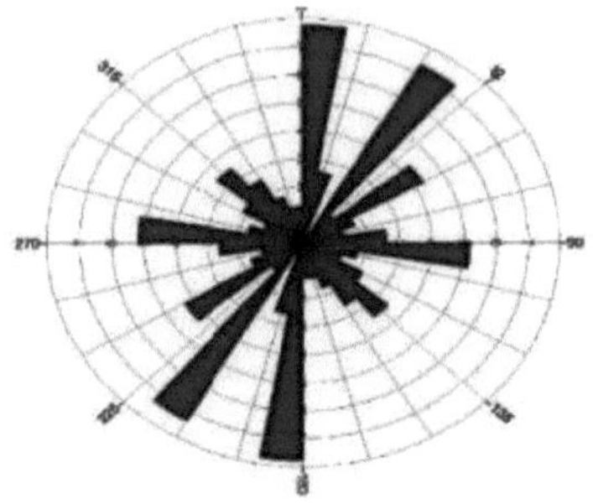

24 September 2018 24 September 2018 24 September 2018

Figure 4.3. Density maps and rose diagrams of lineament structures (continued)

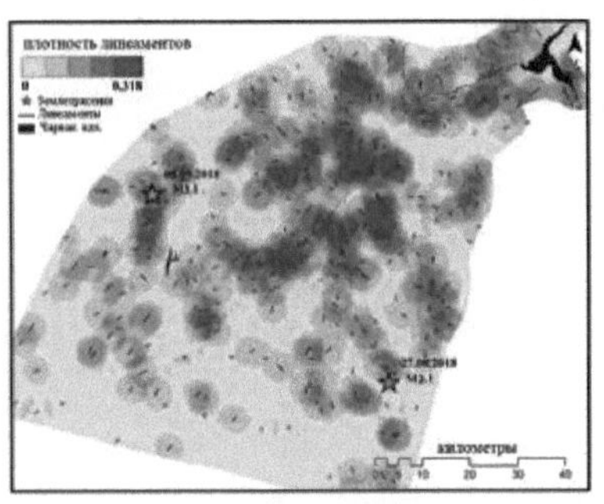

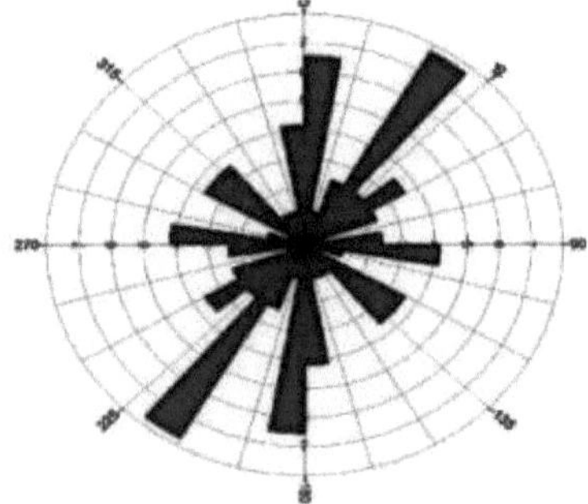

26 October 2018 26 October 2018 26 October 2018

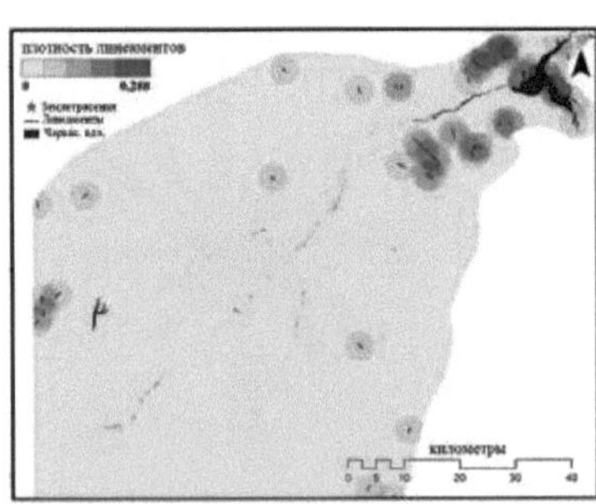

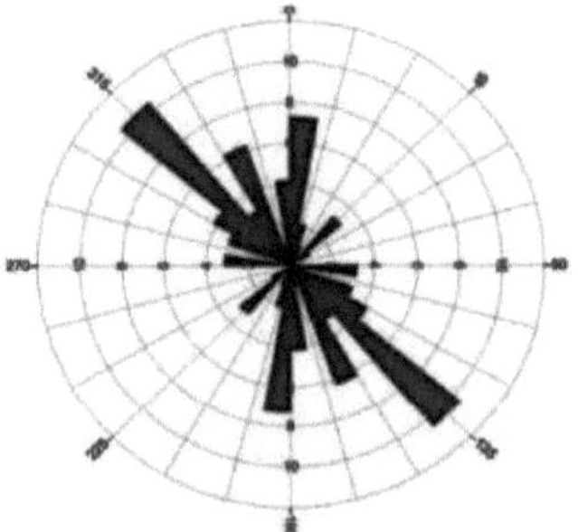

11 November 2018 11 November 2018 11 November 2018

Figure 4.3. Density maps and rose diagrams of lineament structures (continued)

The TGP region is notable for its high seismic activity, encompassing low to medium magnitude earthquakes, with the most significant ones having a magnitude of Mw3.0. In contrast to previous studies conducted by various researchers, where the focus was predominantly on higher magnitude earthquakes [109; pp. 36-47, 111; c. 47-56, 113; c. 3-20, 115; c. 1-26, 118; c. 1283-1291, 130; c. 561-567], this study was the first to consider seismic events with lower magnitude, which is a characteristic feature of this region.

4.2 Study of seismic activity of Almalyk-Angren industrial zone based on lineament analysis

This study aims to use an automated method of lineament extraction from satellite images to study seismic activity in the Almalyk-Angren industrial zone in Uzbekistan. The scientific problem we will address is related to the need for a better understanding of the patterns and characteristics of seismic activity in this region due to its complex tectonic structure. By using a method of automated lineament extraction and comparing the results with geological maps, we hope to efficiently and accurately identify linear features associated with tectonic activity in a large volume of data. This contributes to a better understanding of the geological and tectonic features of the region. We plan to investigate the relationship between density, lineament orientation and seismic activity in the region, and to evaluate the potential of automated lineament analysis as a tool for seismic hazard assessment. Hypotheses are formulated on the relationship between certain lineaments detected by automated analyses and tectonic lines where stress accumulation in the crust can trigger earthquakes. These hypotheses will be tested by analysing data on earthquakes and geological structures, comparing them with the distribution of lineaments. In this way, we will gain a better understanding of the underlying geology and tectonic processes that influence seismic activity in the region. In conclusion, this study aims to investigate lineament dynamics in the region, investigate the localisation of earthquake epicentres based on fault locations, compare the results with geological data and identify areas of anthropogenic influence. The use of automated lineament analysis can improve our understanding of seismic activity and associated potential hazards in the region.

The Angren-Almalyk region is located in the north-eastern part of Uzbekistan and is one of the key ore-bearing areas of the country. The Angrenovo coal mine with a depth of up to 300 metres, the Jirgistan open pit, and the Naugarzan and Apartak coal pits are located here. The Almalyk Mining and Metallurgical Combine (AMMC) is the main producer of of non-ferrous metals in Uzbekistan (Figure *4.4)*.

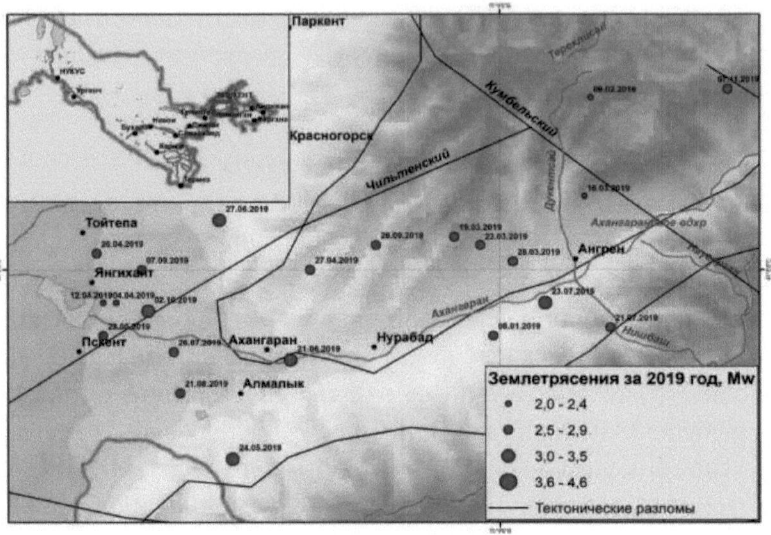

Figure 4.4. Landsat 8 image of the study area, tectonic faults and location of earthquake epicentres [145]

The region is located within the Chatkal-Kuramin block, part of the Tien Shan geological area. One of the main geological structures of this region is a magmatic formation covering about 85% of the territory with the age from Proterozoic to Mesozoic. The structures in the Kuramin zone were formed during the Caledonian, Hercynian and Alpine tectonic cycles. The modern relief of the zone was formed during the Neogene. The north and south of the Angren zone are characterised by tectonic disturbances superimposed on the Chatkal and Kuramin mountains. The tectonics of the region is represented by a complex combination of different structural elements, including folds, troughs, faults, intrusions and volcanic rocks.

The main deposit in the region is the Almalyk ore belt, which stretches over 40 kilometres and is rich in copper, gold, silver and other minerals. The ores are found in carbonate rocks and volcanic deposits formed as a result of volcanic activity and magmatic processes during the Mesozoic era (151). The region is located in the Chatkal-Kuramin block, which is a crystalline massif formed as a result of ancient geological processes that occurred more than 2 billion years ago. The block contains a variety of granites, gneisses, schists and quartzites that have undergone severe tectonic deformation resulting in the formation of various structural elements, including folds, faults and fractures. The most intense tectonic processes in this area occurred in the Mesozoic and Cenozoic, when collision and subduction of lithospheric plates led to the formation of mountain folds, including the Pamir and Tien Shan ranges. These processes resulted in the formation of a variety of rocks, including granites, gneisses, schists and sandstones in the Angren and Almalyk regions. Currently, tectonic processes continue in this area. There are several major faults around the Angren-Almalyk region that define the tectonic structure of this area. One of the longest faults is the Chiltan fault, which extends west of the Angren massif. It is an active fault capable of producing earthquakes. Another significant fault is the Almalyk fault, which runs through the Almalyk massif north of the city of Almalyk. This fault is also active and capable of producing strong earthquakes. According to a study of stress distribution and zoning of the western Tien Shan and adjacent territory based on the excess tangential stress calculated by mathematical modelling and compared with measurements in deep boreholes, this area is identified as a zone with the highest values of excess tangential stress (>2 MPa) [152]. The Kumbel fault, which runs along the southern edge of the Angren-Almalyk zone, is less active but can still produce earthquakes. The fault is a complex shear fault with fault plane inclination angles ranging from 60° to 90° in the northeast, and the southwest block is uplifted. Recently, shear movements with amplitudes up to 5 km have occurred along this fault [6, 153]. The main faults are accompanied by numerous minor auxiliary and associated fractures. Due to intensive mining activities and special geological and tectonic conditions, the Almalyk-Angren industrial zone is a region with a high level of seismic activity.

Synthesised Landsat 8 images combined with a higher quality panchromatic channel were used in the study. The LEFA algorithm was chosen to analyse linear structures, providing automatic extraction of linear objects with high accuracy. A combination of methods such as Canny and Hough Transform was used for image processing and lineament extraction. This approach allowed the study of crustal dynamics at depths of tens of kilometres. Lower resolution images are needed to detect events over larger areas. Images with a resolution of 10-30 metres are useful for studying earthquakes because they can integrate information about the presence of faults at depths of tens of kilometres. Although such images cannot detect individual fractures, they are excellent for tracking changes associated with strength accumulation or weakening caused by tectonic plate movement. In our study, we used Landsat 8 OLI images at 30 m resolution before and after the earthquake downloaded from the United States Geological Survey (USGS) Earth Explorer website [154, 155]. The archived images included data for 6 May, 23 June, 9 and 25 July, 10 August, 11 and 27 September 2019. The catalogue of earthquakes in the region from the Republican Seismic Prognostic Monitoring Centre of the Ministry of Emergency Situations of the Republic of Uzbekistan [24] was used for the analysis. During 2019, 29 earthquakes (2.2 <Mw <4.2) occurred in the Almalyk-Angren industrial zone. Table 4.4 presents the dates selected for analysis (2.5 <Mw <3.2).

Table 4.4.

Catalogue of earthquakes on the territory of Almalyk-Angren industrial zone

for 2019

№	Date	Depth (km)	Magnitude	Epicentre region
1	21.06.2019	5	3	Central Uzbekistan
2	27.06.2019	14	3	Central Uzbekistan
3	21.07.2019	5	2,5	Eastern Uzbekistan
4	23.07.2019	15	3,2	Central Uzbekistan
5	26.07.2019	7	2,9	Central Uzbekistan
6	07.09.2019	5	2,6	Central Uzbekistan
7	26.09.2019	30	2,9	Central Uzbekistan

One of the main statistical measures used to predict lineament changes were rose diagrams and density maps, as shown in Fig. 4.5 (right column). The dominant directions were found to be NW, WNW and NW-SE. It was found that as the number of lineaments increased, the length of the northward direction also increased. The density maps (Figure 4.5, left column) showed high density of lineaments near the epicentre of the earthquakes that occurred on 23 June 2019 and 27 June 2019. The lineament density decreased on 9 May, 9 July and August as indicated by the results with the minimum density recorded on 9 July 2019 equal to 0.33. A high lineament density near the epicentre of the earthquake that occurred on 26 July 2019 was observed on 25 July 2019, the lineaments reached their maximum one day before the event. The maximum number of lineaments was observed on 27 September 2019 with a density of 1.11.

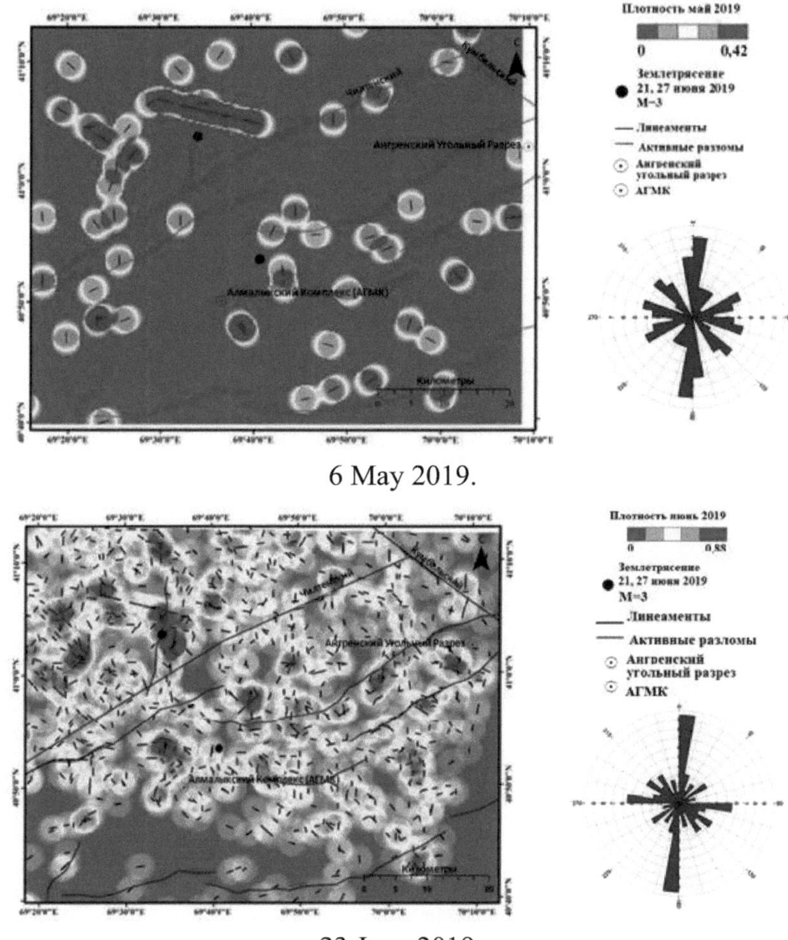

6 May 2019.

23 June 2019.

Fig.4.5 Density maps and lineament diagram roses

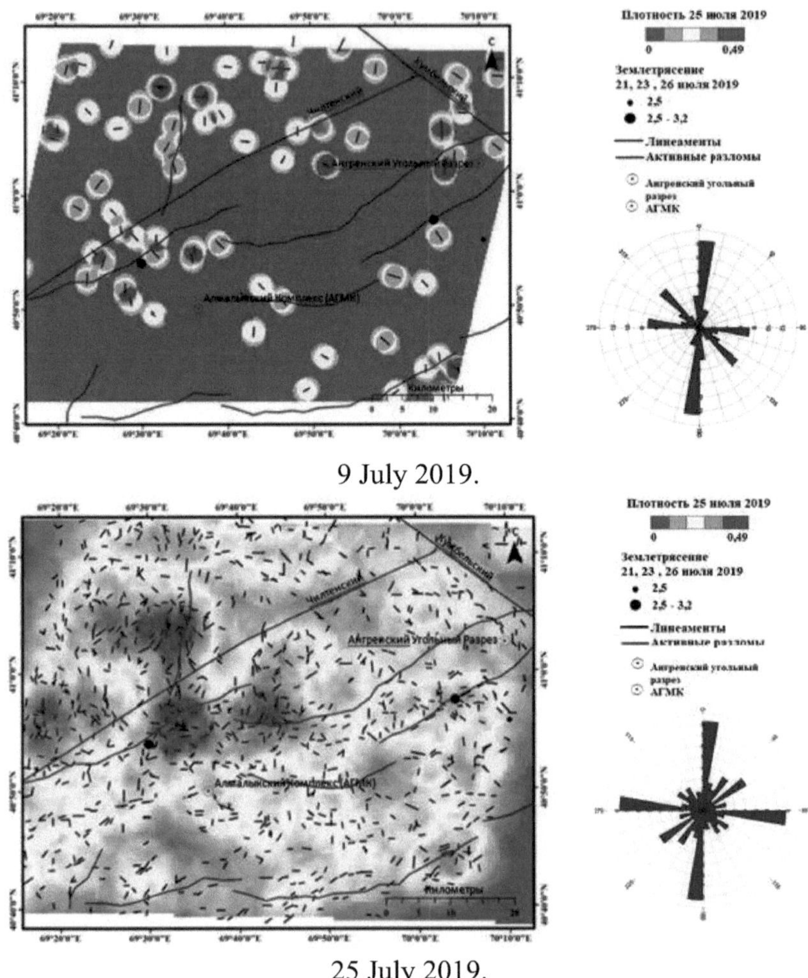

9 July 2019.

25 July 2019.

Fig.4.5 Density maps and lineament diagram roses (continued)

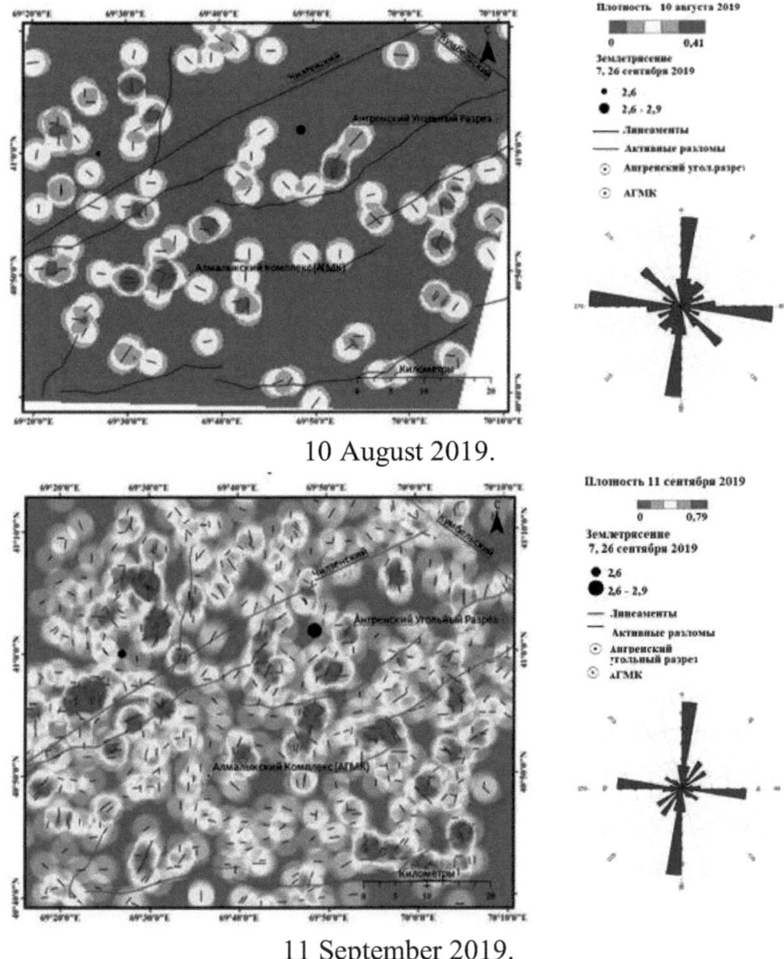

10 August 2019.

11 September 2019.

Fig.4.5 Density maps and lineament diagram roses (continued)

The results showed a clear increasing trend in the number of lineaments in the interval before and after the earthquake, with a sharp decrease in the total number of lineaments 20 days before and approximately 14 days after the earthquake. The pink plots and density maps showed dominant CC, WNW and NW-SE directions, with high lineament densities near the epicentres of the earthquakes that occurred on 23 June 2019, 27 June 2019 and 26 September 2019. The results of the statistical analysis of lineament densities during earthquakes provide important insights into the tectonic processes occurring in the study region and may indicate the possibility of predicting earthquakes in the future. However, these results are only applicable to earthquakes of magnitude around 3 and are concentrated in areas where Paleozoic rocks are exposed at the surface, indicating the influence of human activity on the stress state of the earth's surface in mining or quarrying areas. Further research could extend this work to include analysis of higher magnitude earthquakes in other geological regions as well, and to examine the influence of human activity on tectonic processes. It is also clear from the analysis that there is a tendency for earthquake epicentres to be located along faults. Automated lineament analysis based on remote sensing data can be, one of the most effective methods with speed and efficiency for geodynamic monitoring of seismically hazardous areas, as well as industrial areas. The obtained results showed the possibility of practical application of such analysis, especially for the analysis of fault tectonics, which are zones of increased seismic activity and deformations of the earth surface. The proposed method of studying the dynamics of the lineament system on satellite images, in combination with other methods, can be used for operational monitoring of seismic hazards. It should be noted that the automated method still needs to be improved using mathematical methods and algorithms to make it possible to apply it in areas with different geological conditions [156].

Conclusions to chapter four

This part of the monograph presents the results of experimental studies of the relationship between lineament structures and seismicity. Automatic analysis of lineaments based on remote sensing data demonstrates high efficiency.

The results of seasonal analysis of lineament structures near the technogenic object as Charvak reservoir indicate a minimum number of lineaments in December and a significant increase in March, June and September. The influence of water level fluctuations on lineament formation is observed. Lineament density maps show increased concentrations in the southeast and high densities in the northern part of the region during the winter-spring period, possibly related to seasonal processes such as snowfall. Maps for June and September reveal faults resulting from fluctuations in reservoir levels.

The statistical analysis of the number of lineament structures and selection of the main directions using density maps and rose diagrams for tectonically active territory as Tashkent geodynamic polygon allowed to reveal the dynamics of changes in lineament structures. It was noted that the number and direction of lineaments undergo significant changes about 2-4 months before the earthquake, after which they gradually return to the initial state. Rapid growth of lineament density starts about 20 days before the event, reaches a maximum 4 days before the earthquake and decreases within 16 days after the earthquake.

The analysis showed the effectiveness of the method, especially for earthquakes with magnitude about 3 for the Almalyk-Angren industrial area. The results of statistical analysis indicate an increase in the number of lineaments before and after earthquakes. The tendency of earthquake epicentres to faults is also revealed.

Automated lineament analysis can be an effective method of geodynamic monitoring, but requires additional improvement for application in different geological conditions and zones with different seismic activity. The results obtained showed the possibility of practical application of such analysis, especially for the analysis of tectonic fractures, which are zones of increased seismic activity and deformations of the earth surface. The proposed method of investigating the dynamics of lineament structures using satellite images together with other methods can be used for operational monitoring of seismic hazard. It should be noted that the automatic method still needs to be improved with mathematical methods and algorithms so that it can be applied in the territories with different geological conditions. This is especially true for platform geological areas where, in addition to tectonic faulting, there is a well-developed network of watersheds and other lines. The physical nature of the delineation of these areas must be determined by integrated methods (microwave and GNSS data).

CONCLUSION

The following results were obtained during the study of tectonic processes in the Tashkent region using satellite data. The analysis of the spatial velocity field of 12 GNSS points for the period 20182020 allowed to calculate horizontal velocities in the range of 21-33 mm/y with separation of rotational movements along the tectonic plates.

An improved local geoid model based on SRTM30, ASTER GDEM2 and ALOS AW3D30 data is recommended for geodetic and hydrological studies.

Interpretation of lineament structures based on space images revealed the influence of water level fluctuations in the reservoir on the deformation of the area. Landsat-8 automated lineament analysis confirmed that the cracking process increases several months before the earthquake, reaching a maximum a few days before the event. Statistical analysis highlighted the increase of lineaments before and after earthquakes, as well as the tendency of epicentres to rupture.

The results obtained can be used to assess anthropogenic impacts and to develop a dynamic reference datum for the national geocentric coordinate system.

LIST OF REFERENCES

1. Gatinsky Yu.G., Rundquist D.V. Geodynamics of Eurasia - plate tectonics and block tectonics // Geotectonics. - 2004. -T. 38, № 1. -C. 3-20.
2. Ulomov V.I. On the role of horizontal tectonic movements in seismogeodynamics and seismic hazard prediction // Physics of the Earth. - 2004. - № 9. -C. 14 - 30.
3. Ibragimov R.N., Plotnitsky A.Y., Sadykov Y.M. Seismotectonic conditions of occurrence of the Ghazli earthquakes // Ghazli earthquakes of 1976 and 1984. -Tashkent: Fan, 1986. -C. 18-28.
4. Usmanova, M.T. About geodynamic models and seismicity of Central Asia / M.T. Usmanova // Catalogue of seismic prognostic observations in the territory of Azerbaijan: annual journal. - Baku, 2011. - C. 138-145.
5. Khamidov H.L. Identification of morphokinetic indicators of modern geodynamics of the Western Tien Shan // Modern geodynamics of Central Asia and natural hazards: results of research on a quantitative basis: Proceedings of the All-Russian meeting and youth school on modern geodynamics (Irkutsk, 23-29 September 2012). - Irkutsk: IZK SB RAS, 2012. T. 2. -C. 88-91.
6. Yarmukhamedov A.R. Modern geodynamic activity of the Earth's crust and its connection with seismicity. - Tashkent: University, 1995. - 130 c.
7. Abdurakhmanov R.Z., Demyanov G.V., Kaftan V.I., Pobedinsky G.G. Methodological issues of global and regional geodetic networks construction // Problems and Solutions. - 2013. - №2 (49). - C. 67-70.
8. Abdullabekov, K.N., Maksudov , S.H., Muminov, M.Y., Tuichiev, A.I. On the relationship of geomagnetic field variations with seismic and geodynamic processes in the Earth's crust // Actual problems of modern seismology: Proceedings of the International Conference (Tashkent, 12-14 October 2016). - Tashkent, 2016. - C. 168-171.
9. Khamidov L.A., Ergeshov I.M., Makhkamov S.M., Khamidov H.L. Estimates of modern displacements of the central part of the Fergana Valley // Eighteenth Ural Youth Scientific School on Geophysics: Collection of scientific materials. - Perm: GI Ural Branch of the Russian Academy of Sciences, 2017. - C.228-233.
10. Khusomiddinov S.S., Starovatov A.A., Sadirov F.H. On the role of tidal forces in seismogenic processes // Problems of Seismology. - 2019. -

№2. -C. 29-35.
11. Rebetsky Y.L., Ibragimova T.L., Ibragimov R.S., Mirzaev M.A. Tension state of seismically active areas of Uzbekistan // Voprosy Engineering Seismology. - 2020. - T.47. - №3. - C. 28-52. DOI:10.21455/VIS2020.3-2.
12. Abdrakhmatov, K.Y. et al. Relatively recent construction of the Tien Shan inferred from GPS measurements of present day crustal deformation rates// Nature. -1996, -№384, -P. 450-453.
13. Reigber Ch., Klotz J., Angermann D., Trapeznikov Y. A., Tatevian S. K., Makarov V. I., Abdullabekov K. N., Yuldashbaev T. S., Tsurkov V. G., Kurskeev A. K. Pamir - Tienshan GPS Project: Network, Observation Campaign 92 and Analysis Strategy. In: Montag H., Reigber C. (eds)/ Geodesy and Physics of the Earth. International Association of Geodesy Symposia. Springer, Berlin, Heidelberg. 1993. Vol. 112. P. 42-45.
14. Ergeshov I.M., Makhmudov M.D., Fazilova D.Sh. GPS data processing in GAMIT/GLOBK: on the example of permanent stations of Uzbekistan network // Universum: technical sciences: electron. nauchn. zhurn. - 2020. - 10(79). -C. 50-55. URL: https://7universum.com/ru/tech/archive/item/10817.
15. Herring T.A., Hager B.H., Meade B., Zubovich A.V. Contemporary horizontal and vertical deformation in the Tien Shan // International seminar "On the Use of Space Techniques for Asia-Pacific Regional Crustal Movements Studies". -Moscow, GEOS, 2002, -P. 75-84.
16. Kairanbaeva A.B. Geodetic monitoring and computerised monitoring Modelling of geodynamic processes in the territory of the Northern Tien Shan: Ph. D. . Doctor of Philosophy (PhD). -Alma-Aty, 2014. -145c.
17. Khusomiddinov S. S. Actual issues of seismology in the light of the concept of population and territory protection from emergency situations // Problems of seismology in Uzbekistan. - 2007. -№ 4. - C.10-15.
18. Ergeshov I. M., Khusomidinov A. S., Khamidov X. L. Morphogenetic features of the Eastern part of the Western Tien Shan for the organisation of GPS measuring points // Reports of the Academy of Sciences of the Republic of Uzbekistan. - 2016. - № 3. - C. 49-54.
19. Abdullabekov K.N., Maksudov S.H., Tuichiev A.I., Yusupov V.R. Local variations of the geomagnetic field of anthropogenic and geodynamic nature in the area of the Charvak reservoir // Ecological

Bulletin of Uzbekistan. -2012. -№4. - C.11-15.
20. Abdullabekov K.N., Maksudov S.H., Muminov M.Y., Tuichiev A.I., Geomagnetic studies in the territories of technogenic objects in Uzbekistan. // "Modern problems of geodynamics and geoecology of intracontinental orogens": Proceedings of the Fifth International Symposium (Bishkek, 19-24 June 2011). -Bishkek, 2011. - C.5-8.
21. Yusupov V. Anomalies of geomagnetic field related to natural and technogenic events in Charvak area // Geodesy and Geodynamics. - 2018. - Vol. 9, Issue 5. - P.367-371.
22. Giardini D., Grunthal G., Shedlock K. M. and Zhang P. The GSHAP Global Seismic Hazard Map [In: Lee W., Kanamori H., Jennings P. and Kisslinger C. (eds.)]: International Handbook of Earthquake & Engineering Seismology, International Geophysics Series 81 B, Academic Press, Amsterdam, -2003. - P.1223-1239.
23. Artikov T.U., Ibragimov R.S., Ibragimova T.L., Kuchkarov K.I., Mirzaev M.A. Quantitative assessment of seismic hazard for the territory of Uzbekistan according to the estimated maximum ground oscillation rates and their spectral amplitudes // Geodynamics & Tectonophysics. - 2018. -9(4). -P.1173-1188. doi:10.5800/GT-2018-9-4-0389.
24. Artikov T.U., Ibragimov R.S., Ibragimova T.L, Mirzaev M.A. Complex of general seismic zoning maps OSR-2017 of Uzbekistan // Geodesy and Geodynamics. -2020. -Vol. 11(4). -P. 273-292. https://doi.org/10.1016Zj.geog.2020.03.004.
25. Mavlyanova N. G., Ibragimov R. S., Ibragimova T. L., Rakhmatullaev K. K. Features of Seismogravitational Processes in Zones of Active Earthquake Manifestation in Central Asia: A Case Study of Uzbekistan // Geoecology, Engineering Geology, Hydrogeology, Geocryology. -2021. - Vol. 2. -P. 27-40.
26. Report on research work Geodynamic studies of modern movements of the Earth's crust on geodynamic polygons of the Republic of Uzbekistan by geodetic methods. -Tashkent, 2010. -43 c.
27. Rakhmatullaev Sh., Huneau F., Celle-Jeanton H., Le Coustumer Ph., Motelica- Heino M., Bakiev M. Water reservoirs, irrigation and sedimentation in Central Asia: A first cut assessment for Uzbekistan // Environmental Earth Sciences. - 2013. - Vol.68. - №4. - P. 985-998. doi: 10.1007/s12665-012-1802-0.

28. Fazilova D., Magdiev H. Comparative study of interpolation methods in the development of local geoid // International Journal of Geoinformatics. - 2018. - V.14, №1. -P. 29-33.
29. Fazilova D. Uzbekistan coordinate system transformation from CS42 to WGS84 using deformation grid model // Geodesy and Geodynamics. - 2022. -Vol. 13, Issue 1. - P. 24-30. https://doi.org/10.1016/j.geog.2021.10.001.
30. Fazilova D., Magdiev H. Updating of digital topographic maps in the new national spatial coordinate system: case Fergana valley in Uzbekistan // Proc. Int. Cartogr. Assoc., (Florence, December 14-18, 2021). -Florence, 2021. - Vol.4. -31p. https://doi.org/10.5194/ica-proc-4-31-2021.
31. Fazilova D.Sh., Magdiev X.H. Creation and updating of the altitude basis of topographic maps in the national spatial coordinate system: the example of the Fergana Valley // InterKarto. InterGIS. Geoinformation support of sustainable development of territories: Proceedings of the International Conf. M: Faculty of Geography, Moscow State University, 2021. T. 27. PART 2. P. 155-164. DOI: 10.35595/2414-9179-2021-2-27-155-164.
32. Fazilova D., Ehgamberdiev Sh, Kuzin S. Application of time series modelling to national reference frame realization// Geodesy and Geodynamics, - 2018. -Vol. 10. - Issue 4. -P. 281-287. https://doi.org/10.1016/j.geog.2018.04.003.
33. Hofmann-Wellenhof B., Lichtenegger H., and Collins J. GPS. Theory and Practice. - NY.: Springer-Verlag Wien, 1992. - 326 P.
34. Soudarin L, Cretaux J.F., Cazenave A. Vertical crustal motions from the DORIS space-geodesy system. // Geophys Res Lett. -1999. - №26(9). - P.1207-1210.
35. Herring T.A., King R.W., Floyd M.A., McClusky S.C. Introduction to GAMIT/GLOBK, Release 10.7. - Cambridge.: Department of Earth, Atmospheric, and Planetary Sciences Massachusetts Institute of Technology, 2018. - 54 p.
36. Herring T. A., R. W. King, McClusky S.C. Global Kalman filter VLBI and GPS analysis programme, GLOBK Reference Manual, Release 10.5. - Cambridge.: Department of Earth, Atmospheric, and Planetary Sciences Massachusetts Institute of Technology, 2010b. URL: http: //chandler.mit.edu/~simon/gtgk/GLOBK Ref. pdf

37. Altamimi Z., Rebischung P., Metivier L., Collilieux X. ITRF2014: A new release of the International Terrestrial Reference Frame modelling nonlinear station motions // Journal Geophysical Research: Solid Earth. - 2016. - Vol.121. - №8. - P. 6109-6131. Doi:10.1002/2016JB013098.
38. Zubovich A.V., Wang X.Q., Scherba Y.G., Schelochkov G.G., Reilinger R., Reigber C., Mosienko O.I., Molnar P., Michajljow W., Makarov V.I., Li J., Kuzikov S.I., Herring T.A., Hamburger M.W., Hager B.H., Dang Y., Bragin V.D., Beisenbaev R. GPS velocity field for the Tienshan and surrounding regions // Tectonics. -2010. -Vol. 29. -TC6014. DOI: 10.1029/2010TC002772.
39. Kahle H.G., Cocard M., Peter Y. GPS-derived strain rate field within the boundary zones of the Eurasian, African, and Arabian Plates // J. Geophys. Geophys. Res. Solid Earth. - 2000. -P. 23353-23370.
40. Allmendinger R. Reilinger R., Loveless J. Strain and rotation rate from GPS in Tibet, Anatolia, and the Altiplano // Tectonics. -2007. -V. 26. - Issue 3. - TC3013. DOI: 10.1029/2006TC002030.
41. Smith W. H. F., Wessel P. Gridding with continuous curvature splines intension // Geophysics. -1990. -55 (3). -P. 293-305. DOI: 10.1190/1.1442837.
42. Bogusz J., Kl os A., Grzempowski P., Kontny B. Modelling the velocity field in a regular grid in the area of Poland on the basis of the velocities of European permanent stations // Pure Appl. Geophys. -2014. - 171. -P. 809-833. https://doi.org/10.1007/s00024-013-0645-2.
43. Wessel P., Luis J. F., Uieda L., Scharroo R., Wobbe F., Smith W. H. F., Tian D. The Generic Mapping Tools version 6. // Geochemistry, Geophysics, Geosystems. -2019.-V.20.-P.5556-5564. https://doi.org/10.1029/2019GC008515.
44. Smith W. H. F., Wessel P. Gridding with continuous curvature splines intension // Geophysics. -1990. -55 (3). -P. 293-305. DOI: 10.1190/1.1442837.
45. Hackl M., Malservisi R., Wdowinski S. Strain rate patterns from dense GPS networks // Nat. Hazards Earth Syst. Sci. -2009. -Vol. 9. -P. 1177-1187. https://doi.org/10.5194/nhess-9-1177-2009.
46. Bian W., Wu J., Wu W. Recent crustal deformation based on interpolation of GNSS velocity in continental China // Remote Sens. - 2020. -V. 12. -P.3753. DOI: 10.3390/rs1222223753.

47. Makhmudov M.D., Fazilova D.Sh. Construction the velocity Field in a regular grid in the Tashkent Region on the basis interpolation of GNSS permanent stations data // InterCarto. InterGIS. GI support of sustainable development of territories: Proceedings of the International conference. Moscow: MSU, Faculty of Geography, 2023. V. 29. Part 1. P. 535-545. DOI: 10.35595/2414-91792023-1-29-535-545 (in Russian)
48. Sychugova L.V., Fazilova D.Sh. Analysis of earth surface deformations on the territory of Tashkent region using satellite methods
// Reports of the Academy of Sciences of the Republic of Uzbekistan. - 2022. - № 3. - C.8285.
49. Fazilova D., Sichugova L. Deformation analysis based on GNSS measurements in Tashkent region. // E3S Web of Conferences. - France, 2021. 227, 04002. - P. 1-8. DOI: 10.1051/e3sconf/202122704002.
50. England P., Molnar P. Surface uplift, uplift of rocks, and exhumation of rocks // Geology. - 1990. - Vol.18. - P. 1173-1177.
51. Hutchinson M.F., Gallant J.C. Digital elevation models and representation of terrain shape. In: J.P. Wilson & J.C. Gallant (eds) Terrain Analysis: Principles and Applications. - 2000. P. 29-50. New York, NY: John Wiley & Sons, Inc.
52. Wilson J.P. Environmental applications of digital terrain modelling. NJ: Wiley - Blackwell, 2018. - 336 p.
53. Toz G., Erdogan M. DEM (Digital Elevation Model) production and accuracy modelling of DEMs from 1:35,000 scale aerial photographs // The International Archives of the Photogrammetry, Remote Sensing and Spatial Information Sciences. - Beijing, 2008. - Vol. XXXVII. - Part B1. - P. 775-780.
54. Miliaresis G.C., Paraschou C.V.E.. Vertical accuracy of the SRTM DTED level 1 of Crete // International Journal of Applied Earth Observation and Geoinformation. - 2005. - Vol.7. - No.1. - P. 49-59.
55. Gabriel A., Goldstein R. Crossed-orbit interferometry: theory and experimental results from SIR-B // International Journal of Remote Sensing. - 1988. - Vol.9. - No.5. - P. 857-872.
56. Wechsler S.P. Perceptions by Digital Elevation Model Users of DEM Uncertainty // URISA Journal. - 2003. - Vol.15. - No.2. - P. 61-69.
57. Chen C.F., Wang X., Yan C.Q., Guo B., Liu G.L.. A total error-based multiquadric method for surface modelling of digital elevation models //

GIScience & Remote Sensing. - 2016. - Vol.53. - №5. - P. 578-95.
58. Fan L., Smethurst J., Atkinson P., Powrie W. Propagation of vertical and horizontal source data errors into a TIN with linear interpolation // International Journal of Geographical Information Science. - 2014. - Vol.28. - No.7. - P. 1378-400.
59. Zink M., Geudtner D. Calibration of the interferometric X-SAR system on SRTM // In: Proceedings of the CEOS SAR workshop. (Toulouse, 26-29 October France, 1999), -1999. -P.375-377.
60. Gonzalez-Moradas M., Viveen W. Evaluation of ASTER GDEM2, SRTMv3.0, ALOS AW3D30 and TanDEM-X DEMs for the Peruvian Andes against highly accurate GNSS ground control points and geomorphological-hydrological metrics // Remote Sensing of Environment. - 2020. - Vol.237. doi: 10.1016/j.rse.2019.111509.
61. Hawker L., Neal J., Bates P. Accuracy assessment of the TanDEM-X 90 Digital Elevation Model for selected floodplain sites // Remote Sensing of Environment. - 2019. - Vol.232. doi: 10.1016/j.rse.2019.111319.
62. Liu K., Song C., Ke L., Jiang L., Pa, Y., Ma R. Global open-access DEM performances in Earth's most rugged region High Mountain Asia: A multi-level assessment // Geomorphology. - 2019. - Vol.338. - P. 16-26. doi: 10.1016/j.geomorph.2019.04.012.
63. Baugh C. A., Bates P. D., Schumann G., Trigg M. A. SRTM vegetation removal and hydrodynamic modelling accuracy // Water Resources Research. - 2013. - Vol.49. - №9. - P. 5276-5289. doi: 10.1002/wrcr.20412.
64. Robinson N., Regetz J., Guralnick R.P. EarthEnv-DEM90: a nearly-global, void-free, multi-scale smoothed, 90m digital elevation model from fused ASTER and SRTM data // ISPRS Journal Photogrammetry Remote Sensing. - 2014. - Vol.87. - P. 57-67. doi: 10.1016/j.isprsjprs.2013.11.002.
65. O'Loughlin F.E., Paiva R.C., Durand M., Alsdorf D., Bates P.. A multi-sensor approach towards a global vegetation corrected SRTM DEM product // Remote Sensing of Environment. - 2016. - Vol.182. - P. 49-59. doi: 10.1016/j.rse.2016.04.018.
66. Yamazaki D., Ikeshima D., Tawatari R., Yamaguchi T., O'Loughlin F. Neal J., et al. A high-accuracy map of global terrain elevations // Geophysical Research
Letters. - 2017. - Vol.44. - No.11. - P. 5844-5853. doi:

10.1002/2017GL072874.
67. Yue L., Shen H., Zhang L., Zheng X., Zhang F., Yuan Q. High-quality seamless DEM generation blending SRTM-1, ASTER GDEM v2 and ICESat/GLAS observations // ISPRS Journal of Photogrammetry and Remote Sensing. - 2017. - Vol.123. - P. 20-34. doi: 10.1016/j.isprsjprs.2016.11.002.
68. Schumann G.J., Bates P.D., Neal J.C., Andreadis K.M. Fight floods on a global scale // Nature. - 2014. - Vol.507. - P. 169. doi: 10.1038/507169e.
69. Kenward T., Lettenmaier D.P., Wood E.F., Fielding E. Effects of Digital Elevation Model Accuracy on Hydrologic Predictions // Remote Sensing of Environment. - 2000. - Vol.74. - №3. - P. 432-444. doi: 10.1016/S0034- 4257(00)00136-X.
70. Sanders B. F. Evaluation of on-line DEMs for flood inundation modelling // Advances in Water Resources. - 2007. - Vol.30. - №8. - P. 1831-1843. doi: 10.1016/j.advwatres.2007.02.005.
71. Mukherjee S., Joshi P. K., Mukherjee S., Ghosh A., Garg R. D., Mukhopadhyay A. Evaluation of vertical accuracy of open source digital elevation model (DEM) // International Journal of Applied Earth Observation and Geoinformation. -2013 .-Vol .21. - P. 205-217. doi: 10.1016/j.jag.2012.09.004.
72. Elkhrachy I. Vertical accuracy assessment for SRTM and ASTER Digital Elevation Models: A case study of Najran city, Saudi Arabia // Ain Shams Engineering Journal. - 2018. - Vol.9. - No.4. - P. 1807-1817. doi: 10.1016/j.asej.2017.01.007
73. Khasanov Kh. Evaluation of ASTER DEM and SRTM DEM Data for Determining the Area and Volume of the Water Reservoir // IOP Conf. Series: Materials Science and Engineering. - 2020. - Vol.883 doi:10.1088/1757- 899X/883/1/012063.
74. Zumberge J. F., Heflin M. B., Jefferson D. C. C., Watkins M. M., Webb F. H. Precise point positioning for the efficient and robust analysis of GPS data from large networks // Journal Geophysical Research. - 1997. - Vol.102. - №B3. - P. 5005-5017.
75. Bertiger W., Bar-Sever Y., Dorsey A., Haines B., Harvey N., Hemberger D., Heflin M., Lu W., Miller M., Moore A.W., Murphy D., Ries P., Romans L., Sibois A., Sibthorpe A., Szilagyi B., Vallisneri M., Willis P. GipsyX/RTGx, a new tool set for space geodetic operations and

research // Advances in Space Research. - 2020. - Vol.66. - P. 469-489. Doi: 10.1016/j.asr.2020.04.015.

76. Wolf R. P., Ghilani D. C. Adjustment computations: statistics and least squares in surveying and GIS. - NY.: Wiley, 1997. - 347 p.

77. Deniz I., Ozener H. Estimation of strain accumulation of densification network in Northern Marmara Region, Turkey // Natural Hazards and Earth System Science. - 2010. - Vol. 10. - №10. - P. 2135-2143. Doi: 10.5194/nhess-10- 2135-2010.

78. Barthelmes F., Kohler W. International Centre for Global Earth Models (ICGEM), in: Drewes H., Kuglitsch F., Adam J., et al. // The Geodesists Handbook, Journal of Geodesy. - 2016. - Vol.90. - №10. - P. 907-1205.

79. Peprah M.S., Ziggah Y.Y., Yakubu I. Performance evaluation of the Earth Gravitational Model 2008 (EGM2008) - a case study // South African Journal of Geomatics. -2017.- Vol. 6-№1 .- P. 47-72. https://doi.org/10.4314/sajg.v6i1.4.

80. Fazilova D., Magdiev H., Halimov B. High-precision satellite leveling and investigation of the local geoid model in the territory of Kashkadarya region // EPRA International Journal of Economic Growth and Environmental Issues. - 2020. - Vol.8. - No.4. - P. 31-35. Doi:10.36713/EPRA5788.

81. Fazilova D., Arabov O. Vertical accuracy evaluation of free access digital elevation models (DEMs): case Fergana Valley in Uzbekistan. Earth Sciences Research Journal. -2023.-27(2),- P.85-91. https://doi.org/10.15446/esrj.v27n2.103801.

82. Balenovic I., Marjanovic H., Vuletic D., Paladinic E., Ostrogovic M., Indir K. Quality assessment of high density digital surface model over different land cover classes // Periodicum biologorum. - 2015. - Vol. 117. - № 4. - P. 459470.

83. Zhilin L., Qing Z., Gold C. Digital terrain modelling: principles and methodology. - Fl.: CRC Press, Boca Raton, 2005. - 318 p.

84. Wechsler S. P. Perceptions of Digital Elevation Model Uncertainty by DEM Users // URISA Journal. - 2003. - Vol.15. - P. 57-64.

85. Tachikawa T., Kaku M., Iwasaki A., et al. ASTER Global Digital Elevation Model Version 2 - Summary of Validation Results // by the ASTER GDEM Validation Team. - 2011. - P.21.

86. Fazilova D., Magdiev K., Sichugova L. Vertical Accuracy Assessment

of Open Access Digital Elevation Models Using GPS. // International Journal of Geoinformatics. - Thailand, 2021. - Vol. 17, № 1. - P. 19-26. https://journals.sfu.ca/ijg/index.php/journal/article/view/1701.

87. Hodgson R.A. Classification of structures on joint surface // American Journal of Science. - 1961. - Vol.259. - №7. - P. 493-502. DOI:10.2475/ajs.259.7.493.

88. Nickelsen R.P. New basement tectonics evaluated in Salt Lake City // Geotimes. - 1975. - Vol.20. - №10. - P. 16-17.

89. Hodgson R.A. Review of significant early studies in lineament tectonics // Proceeding of first international conference on the New Basement Tectonics, Salt Lake City, Utah: Utah Geology Associates Publication. - 1974. - №5. - P. 1-10.

90. Kutina J., Hildenbrand T.G. Ore deposits of the western United States in relation to mass distribution in the crust and mantle // Geological Society of America Bulletin. - 1987. - Vol.99. - №1. - P. 30-41.

91. Mollard J.D. Fracture lineament research and applications on the western Canadian plains // Canadian Geotechnical Journal. - 1988. - Vol.25. - P. 749767.

92. Kocharyan G.G. Geomechanics of faults. - M.: GEOS, 2016. - 424 c.

93. Korchuganova N.I., Korsakov A.K. Remote sensing methods of geological mapping. - M.: KDU, 2009. - 288 c.

94. Who and how broke the Earth, or where did the planetary mountain ranges and faults arise from / [Electronic resource]. - Access mode: httpshttps://habr.com/ru/post/565966/habr.com/ru/post/565966/ (date of reference: 22.02.2022).

95. Ali E.A., El Khidir S.O., Babikir A.A., Abdelrahman E.M. Landsat ETM+7 Digital Image Processing Techniques for Lithological and Structural Lineament Enhancement: Case Study Around Abidiya Area, Sudan // The Open Remote Sensing Journal. - 2012. - Vol.5. - P. 83-89.

96. Al-Nahmi F., Alamib O.B., Baiddera L., Khanbaric K., Rhinanea H., Hilalia A. Using remote sensing for lineament extraction in Al Maghrabah area - Hajjah, Yemen // The International Archives of the Photogrammetry, Remote Sensing and Spatial Information Sciences. - 2016. - Vol. XLII-2/W1. - P. 137-142. doi:10.5194/isprs-archives-XLII-2-W1-137-2016.

97. Bondur V.G., Zverev A.T. Physical nature of lineaments registered on space images during monitoring of earthquake-prone territories // Modern

problems of Earth remote sensing from space. - 2006. - B.3. - VOL. 2. - P. P. 177-183.
98. Mallast U., Gloaguen R., Geyer S., Rodiger T., and Siebert C. Derivation of groundwater flow-paths based on semi-automatic extraction of lineaments from remote sensing data // Hydrology and Earth System Sciences. - 2011. - Vol. 15. - P. 2665-2678. DOI: 10.5194/hess-15-2665-2011.
99. Hermi S.O., Elsheikh R.F.A., Aziz M., Bouaziz S. Structural Interpretation of Lineaments Uses Satellite Images Processing: A Case Study in North-Eastern Tunisia // Journal of Geographic Information System. - 2017. - Vol.9. - No.4. - P. 440-455. doi:10.4236/jgis.2017.94027.
100. Bondur V.G., Krapivin V.F., Savinykh V.P. Monitoring and forecasting of natural disasters. - Moscow: Nauch. Mir, 2009. - 692 c.
101. Chaabouni R., Bouaziz S., Peresson H., Wolfgang J. Lineament analysis of South Jene in Area (Southern Tunisia) using remote sensing data and geographic information system // The Egyptian Journal of Remote Sensing and Space Sciences. - 2012. - Vol.15. - P. 197-206.
102. Kassou A., Essahlaoui A. and Aissa M. Extraction of Structural Lineaments from Satellite Images Landsat7 ETM+ of Tighza Mining District (Central Morocco) // Research Journal of Earth Sciences. - 2012. - Vol.4. - No. 2. - P. 44-48.
103. Takorabt M., Toubal A.C., Haddoum H. et al. Determining the role of lineaments in underground hydrodynamics using Landsat 7 ETM+ data, case of the Chott El Gharbi Basin (western Algeria) // Arabian Journal Geosciences. - 2018. - Vol.11. doi: 10.1007/s12517-018-3412-y.
104. Gubin V. H. Satellite technologies in geodynamics. - Minsk: Minsktipproekt, 2010. - 87 c.
105. Prabhakaran A., Jawahar Raj N. Mapping and analysis of tectonic lineaments of Pachamalai hills, Tamil Nadu, India using geospatial technology // Geology, Ecology, and Landscapes. - 2018. - Vol.2. - No. 2. - P. 81-103. doi: 10.1080/24749508.2018.1452481.
106. Hein B.E., Mikhailov A.E. General geotectonics // - M.: Nedra, 1985. -326 c.
107. Marghany M., Hashim M. Lineament mapping using multispectral remote sensing satellite data // International Journal Physical Sciences. - 2010. - Vol.5. - No.10. - P. 1501-1507.

108. Maged M., Mansor S., Hashim M. Geological mapping of United Arab Emirates using multispectral remotely sensed data // American Journal of Engineering and Applied Sciences. - 2009. - № 2. - P.476-480.
109. Gaponova E.B., Zverev A.T., Tsidilina M.H. Identification of lineament anomalies from space images during strong earthquakes in California with magnitudes 6.4 and 7.1 // Earth Exploration from Space. - 2019. - №6. - C. 36-47.
110. Mallast U., Gloaguen R., Geyer S., Reodiger T., Siebert C. Derivation of groundwater flow-paths based on semi-automatic extraction of lineaments from remote sensing data // Hydrology and Earth System Sciences. - 2011. - Vol.15. - P. 2665-2678.
111. Bondur, V.G.; Zverev, A.T. Mechanisms of the lineament formation registered on the space images during the monitoring of the earthquake-prone territories (in Russian) // Earth exploration from space. - 2007. - Vol.1. - P. 47-56.
112. Kritsuk S.G., Latypov I.Sh. Lineament analysis of MODIS data and possibilities of interpretation of its results // Modern problems of Earth remote sensing from space. - 2019. - T.16. - №4. - C. 45-53. doi: 10.21046/2070-7401-2019-16-4-45-53.
113. Bondur V.G., Zverev A.T., Gaponova E.V., Zima A.L. Space research of precursor cyclicity in earthquake preparation manifested in the dynamics of lineament systems // Earth Exploration from Space. - 2012. - №1. - C. 3-20.
114. Zoran M. Earthquake Precursors Assessment in Vrancea Area, Romania by Satellite and Geophysical In-Situ Data // Proc. 'Fringe 2009 Workshop' (Frascati, Italy, 30 November - 4 December 2009). 'Fringe 2009 Workshop' (Frascati, Italy, 30 November - 4 December 2009). - Italy, 2010. - P. 677.
115. Nath B., Niu Z., Acharjee S., Qiao H. Monitoring the geodynamic behaviour of earthquake using Landsat 8-OLI time series data: the case of Gorkha and Imphal // Natural Hazards and Earth System Sciences Discussions. - 2017. - P. 1-26. doi: 10.5194/nhess-2017-10.
116. Zakharov V.N., Zverev A.V., Zverev A.T., Malinnikov V.A., Malinnikova O.N.. Application of automated lineament analysis of satellite images in modern geodynamics research: a case study // Russian Journal Earth Sciences. - 2017. - Vol.17. - P.1-14. doi: 10.2205/2017es000599.

117. Elmahdy S.I., Mohamed M.M. Mapping of tecto-lineaments and investigate their association with earthquakes in Egypt: a hybrid approach using remote sensing data // Geomatics, Natural Hazards Risk. - 2015. - Vol.7. - P. 600-619. doi: 10.1080/19475705.2014.996612.
118. Bondur V.G., Zverev A.T., Gaponova E.V. Precursor variability of lineament systems detected using satellite images during strong earthquakes // Izvestiya,
Atmospheric and Oceanic Physics. - 2019. - Vol.55. - P. 1283-1291. doi: 10.1134/s0001433819090123
119. Kocharyan G.G. Geomechanics of faults. - M.: GEOS, 2016. - 424 c.
120. Stepanov A.V. Field stage of seismic data acquisition:
Training manual for advanced training courses "Petrophysics and geophysics in petroleum geology". - Kazan: Kazan University, 2013. - 35 c.
121. Arseniev-Obraztsov S.S., Pozdnyakov A.P. Application of the InSAR satellite radar interferometry method for solving problems of field geology and oil and gas fields development // Geology and field development. Gas industry. - 2020. - №3. - C. 38-44.
122. Tronin A. A. Space methods of earthquake research.
Current state and prospects // Modern problems of Earth remote sensing from space. - 2004. - B.1. - V.1. - P. 33-38.
123. Roeloffs E., Quilty E. Water level and strain changes preceding and following the August 4, 1985 Kettleman Hills, California, earthquake // Pure and Applied Geophysics. - 1997. - Vol.149. - P. 21-60. doi:10.1007/BF00945160
124. Fraser-Smith A.C., Bernardi A., McGill P.R., Ladd M.E., Helliwell R.A. & Villard O.G. Low-frequency magnetic field measurements near the epicentre of the Ms 7.1 Loma Prieta earthquake // Geophysical Research Letters. - 1990. - Vol.17. - No.9. - P. 1465-1468. doi:10.1029/GL017i009p01465.
125. Sato H. Precursory Land Tilt prior to the Tonankai Earthquake of 1944 // Some Precursors prior to Recent Great Earthquakes along the Nankai Trough. - 1977. - Vol.25. - P. 115-121.
126. Mogi K. Temporal variation of crustal deformation during the days preceding a thrust-type great earthquake. The 1944 Tonankai earthquake of magnitude 8.1 // Pure and Applied Geophysics. - 1984. - Vol.122. - P. 765-

780.
127. Tsunogai U. & Wakita H. Precursory chemical changes in ground water: Kobe earthquake, Japan // Science. - 1995. - Vol.269. - No.5220. - P. 61-63. doi: 10.1126/science.269.5220.61
128. Wakita H. Earthquake chemistry II, collected papers. Laboratory for Earthquake Chemistry, Faculty of Science, University of Tokyo, Tokyo, - 1996. - Vol.II.
129. De Swaaf, Kirt. Da rumort es standig im Untergrund, Interview with Pier Francesco Biagi // Der Standard. - 2011. - 22 Marz. (English).
130. Bondur V.G. and Zverev A.T. A method of earthquake forecast based on the lineament analysis of satellite images // Doklady Earth Sciences. - 2005. - Vol.402. - №4. - P. 561-567.
131. Peresan A., Gorshkov A., Soloviev A., and Panza G. F. The contribution of pattern recognition of seismic and morphostructural data to seismic hazard assessment // Bolletino di Geofisica Teorica ed Applicata. - 2015. - Vol.56. - №2. - P. 295-328. doi:10.4430/bgta0141
132. Arellano-Baeza A.A., Garcia R.V. and Trejo-Soto M. Use of high resolution satellite images for tracking of changes in the lineament structure caused by 2007 earthquakes. - 2007. http://arxiv.org/abs/0706.1827v2
133. Singh V.P. and Singh R.P. Changes in stress pattern around epicentral region of Bhuj earthquake of 26 January 2001 // Geophysica Research Letters. - 2005. - Vol.32. doi:10.1029/2005GL023912.
134. Shevyryov, S.L. LEFA programme: automated structural analysis of the space base in Matlab environment // Uspekhi sovremennoi nauchnostvoznaniya. - 2018. - №10. - C. 138-143.
135. On measures to implement the investment project "Creation of the National Geographic Information System". Presidential Decree No. 2045 of 2013. [Electronic resource]. - Mode of access: http://http://lex.uz/pages/getpage.aspx?lact_id=2242710.uz/pages/getpage. aspx?lact_id=2242710 (date of circulation: 24.09.21)
136. Zlatopolsky A.A. Programme LESSA (Lineament Extraction and Stripe Statistical Analysis) automated linear image features analysis-experimental results // Computers & Geoscience. - 1992. - Vol.18. - No.9. - P. 1121-1126.
137. Duda R.O., Hart P.E.. Use of the Hough Transformation to Detect

Lines and Curves in ENVI - Image Processing and Analysis Solution / [Electronic resource]. - Access mode: https://www.ittvis.com/ envi/ (date of access: 05.05.2021).
138. Geomatica 2013. OrthoEngine. Course Exercises. Version 2013. Richmond Hills, Ontario, Canada: PCI Geomatics Enterprises Inc., - 2013. - 169 p.
139. ArcGIS Resources. Introduction to ArcGIS / [Electronic resource]. - Access mode:httpshttps://resources.arcgis.com/ru/help/gettiresources.arcgis.com/ru/help/getti ng-started/articles/026n00000014000000.htm (accessed 05.05.2021).
140. RockWorks/LogPlot. Ideal tool for modelling / [Electronic resource]. - Access mode : http:// mapinfo.ru/product/rockware (access date: 05.05.2021). Pictures // Comm. ACM. - 1972. - Vol.15. - P. 11-15.
141. ENVI User's Guide. ENVI Version 4.1. Research Systems, Inc. - September 2004. - P. 1111.
142. Chunzhong N., Shitao Z., Chunxue L., Yongfeng Y., Yujian L. Lineament Length and Density Analyses Based on the Segment Tracing Algorithm: A Case Study of the Gaosong Field in Gejiu Tin Mine, China // Mathematical Problems in Engineering. - 2016. - Vol.2016. - Article ID 5392453. - 7 pages. https://doi.org/10.1155/2016/5392453
143. Lillesand T.M., Kiefer R.W., Chipman J.W.. Remote Sensing and Image Interpretation. 5th edition. NY: John Wiley & Sons Inc., 2004. - 666 p. Doi: 10.2307/634969
144. Sichugova L., Fazilova D. Structural interpretation of lineaments using satellite image processing: a case study in the vicinity of the Charvak reservoir // InterCarto. InterGIS. - Moscow: Moscow University Press, MSU, 2020. - V.26, Part 2. - pp. 436-442. DOI: 10.35595/2414-9179-2020-2-26-436-442.
145. Republican Centre of Sesmoprognostic Monitoring of the Ministry of Emergency Situations of RUz // Consolidated Catalogue. [Electronic resource]. - Access mode: URL: https: //rcsm.fvv.uz/ru/catalog_col
146. Shevyrev S. Neotectonics, remote sensing and erosion cut of ore-controlling structures of the Mnogovershinnoe gold-silver deposit (Khabarovsk Krai, Russian Far East) // Ore Geology Reviews. - 2019. -

Vol.108. - P. 8-22. Doi: j.oregeorev.2018.11.016.
147. Shevyrev S. LEFA: Lineament Extraction and Fracture Analysis, 2018. http://lefa.geologov.net.
148. Canny J. A computational approach to edge detection // IEEE Transactions on Pattern Analysis and Machine Intelligence. - 1986. - Vol. PAMI-8. - №6. - P. 679-698. doi: 10.1109/TPAMI.1986.4767851.
149. Ulomov V.I. Dynamics of the Earth's crust in Central Asia and earthquake prediction. - T.: FAN, 1974. - 124 p.
150. Sichugova L. and Fazilova D. The lineaments as one of the precursors of earthquakes: A case study of Tashkent geodynamical polygon in Uzbekistan. // Geodesy and Geodynamics. - Elsevier: China, 2021, - V. 12, No. 6, - pp. 383388. DOI: 10.1016/j.geog.2021.08.002.
151. Mamadjanov Y., Aminov J.X., Hodzhiev A.K., Khalimov G.A. Late Paleozoic shoshonite-latite-monzonitoid magmatism
Chatkal-Kuramin zone of the Middle Tien Shan: geology, petrogeochemistry and potential ore-bearing // Actual problems of geology, geophysics and metallogeny: Proceedings of scientific and technical conference (Tashkent, 11-12 September 2017). -Tashkent, 2017. - C. 46-49.
152. Bakiev M.H., Khamidov L.A., Ibragimov A.H. Stress concentration near local inhomogeneities of the Earth's crust // // Inland Earthquake. China. - 2001. - Vol. 15, № 4 - C. 376-384.
153. Khamidov L.A. Study of stresses fields of Chatkal's mountain zone of West Tien Shan // Geodinamika. -2010. -1(9). -P.57-66.
154. United States Geological Survey (USGS) Earth Explorer website / [Electronic resource]. - Access mode: URL: https://earthexplorer.usgs.gov (date of reference: 23.08.2022).
155. Landsat 8-9 Operational Land Imager (OLI) and Thermal Infrared Sensor (TIRS) / [Electronic resource]. - Access mode: URL: https://www.usgs.gov/faqs/what-are-band-designations-landsat-satellites (date
circulation: 23.08.2022).
156. Sichugova L, Fazilova D. Study of the seismic activity of the Almalyk-Angren industrial zone based on lineament analysis // International Journal of Engineering and Geosciences. -2024. Accepted.
157.

Printed by Books on Demand GmbH, Norderstedt / Germany